The Pool Manager

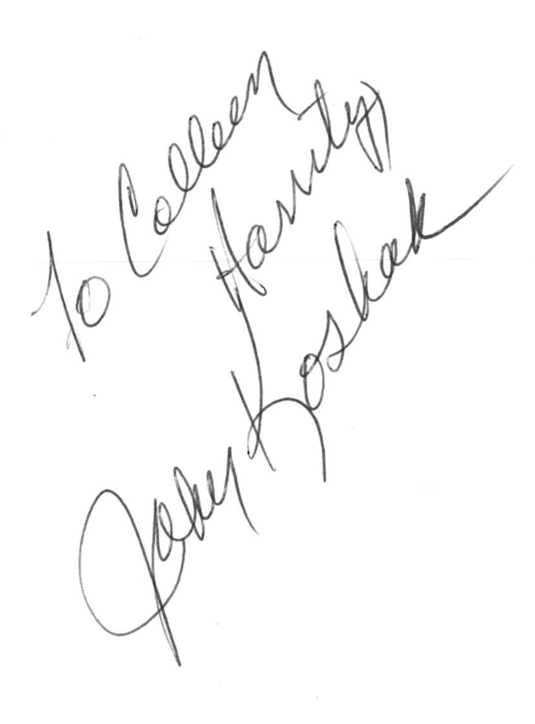

The Pool Manager

John W. Koshak

ISBN 1-4196-2164-5

Library of Congress Control Number: 2005910574

To order additional copies, please contact us.
BookSurge, LLC
www.booksurge.com
1-866-308-6235
orders@booksurge.com

The Pool Manager

Dedicated To Stephanie, Jeremy, Sarah, Shelby, And Austin

I

Karen Jordan looked out her favorite window with a warm smile she only recalled having when pregnant and in a contented state of mind. In the beginning of the second trimester of her third pregnancy, she recalled the warm glow she experienced with her first two. She felt the new life, her bodily resources being shared. Her quirky diet and warm flush feeling at times. Talks with her friends revealed they also had a glow, a warmth that was consistent with having a child with them as well. An interesting effect, probably one precipitating the long-term memory to begin the forgetting process of childbirth. So clever the body is she thought. She thought of the joy she felt with her first pregnancies. Now she was totally aware of what to do and expect; she was a self confident mom.

Karen was not particularly religious, in fact probably quite the opposite. Her work in the field of psychology demonstrated her more pragmatic belief that the organism, homo sapien, operated on a system of rules, some known, some as yet not. Her particular research was in brain energies, which could be manifested in forms of telekinesis to exorcism, from normal mental behavior to serious neurosis. She dedicated the latest part of her career to research; to discover if there was an as yet undiscovered force or energy at work which could explain brain function and behavior. Even her behavior as a mother during her child's maturation in the womb.

But right this minute, her research was as far away as the moon rising over her Julian hillside home appeared to be. She had just awakened, feeling the life inside her grow, feeling the majesty and wonder of it while looking out at the moonrise through her office window. Her life, with all the bucolic peacefulness that a small town can provide, was bliss. Her husband JJ was out of town again, but his daily calls were enough for them to stay connected, to feel secure and loved.

Jeremy Jordan had a unique business life. Always on the road, his days were spent rushing to and from business meetings. The meetings ranged throughout the world. He was a consultant for a large American financial institution. His participation with the International Standards Organization or ISO was dedicated to harmonizing banking and business standards that would assure economic consistency as the worlds businesses were entering global markets. His passion for the work kept him away for sometimes weeks at a time, but Karen busied herself with a life of research and motherhood in their small home town in Southern California. Each of them had professional lives which kept them from finding time to be lonely, but their love for each other was also boundless.

Their two children were both boys, Jeremy and Stephan, five and three, and were prone to getting into as much trouble as young boys can possibly be in; but with the sweetest disposition to add confusion to the results. As they grew old enough to be somewhat independent, Stephan would attempt to keep up with Jeremy. Jeremy's impatience at Stephan's yet developed abilities led to many conversations of patience and helping from Karen. Jeremy was often patient until his latest LEGO model was ruined by Stephan's curiosity or when the hamster was missing from the cage.

Unfettered by the city life, Julian is a small mountain town with miles of forests tucked into mountain valleys that the boys could feel was their playground. Indeed it was, and their trips into nature with JJ and Karen were always interesting and exciting for the boys. Their favorite game in their own real back yard was to bury themselves in the mountainous piles of pine needles and oak leaves which accumulated every day. They had only known Julian, unlike a lot of kids today, they had never moved. They were often outside, hiking, and climbing trees; both were healthy and mostly smiles; with the occasional black eyes that good play can produce. TV was not available to them because they were so remote and JJ felt if they never knew what a TV was, that would be OK with him.

They were not in school, but were probably ready for entry into second grade with the learning environment they lived in. Stephan was younger but did not let that get in the way of his desire to keep up with Jeremy. Even at their young ages, they had a full command of the language and both impressed and astonished Karen and JJ with their choice of words and conceptualization. Lately, they were both quite impressed with collecting bugs in baby food jars. They collected moths, ticks, beetles, ants, even a tarantula or two. Most of their play time was spent with Suzanne, their au pair.

Suzanne kept them occupied and under observation as Karen continued her research on the second story of their home. Suzanne was a godsend. She had many talents, including her love of children and sisterly affection that made the boys pull her into their lives. She also helped in very significant ways with Karen's research. When the kids were asleep, she would be found with Karen in her office discussing the electronics gear, optional strategies and of course the endless stories of the kids antics.

Karen's office faced the southeast, overlooking Lake Cuyamaca which she took dibs on when they moved in. Her love of discovery included her desire to be able to see every sun, moon, and star rise she possibly could. Her favorite time of day was the early morning when the clear mountain air allowed her to see the rising of the Cancer or the moon while sipping her first cup of coffee. In the summer, when Orion dashed upwards from the lake, protecting his celestial love, Karen would sit and dream of many things. Dreams of her love of family; fantasies of what the future held with three healthy children, and how best to approach her work. Trying to discover this force or energy which must be locked away in the human body somewhere and endless writing to secure her scientific results for the eventual paper she would write.

Somewhere there had to be something in the body that

appears at conception and disappears at death. She knew that some referred to it as the soul, but soul or not, whatever it was, could it be isolated? If so, could it be experimented with and shown to be the cause of the numerous forms of aberrant behavior? Could it explain the dark mental conditions that some believed were demons which inhabited the soul needing exorcisms? If the source could be defined, norms could be established and therapeutic solutions could be developed.

Her early residency was at Laguna Mental Hospital in San Francisco. It was the most tragic and interesting time of her life. Always challenging, she discovered she had a deep sense, a passionate desire of wanting to know how and why these diseases existed. Finally, with her residency completed and several years of successful private practice, her desire to research these conditions grew and finally took over. The passion she held for her career launched her to try to discover the energy that had eluded everyone. She never read anything in the medical journals approaching what she was intuitively aware of. She always felt the medicine practiced both in public service and private practice based on the medicine still taught in medical school was limited. There had to be more than just a diagnosis and drug therapies to stabilize patients. She wanted to see these tragic people restored to sanity, their families rejoined and to end the non-traumatic injuries. What else in the power of the brain was being overlooked, she wondered, by the best science had to offer?

For years she had seen many patients given lifelong drug therapies which required constant monitoring. Episodes of ineffective treatment leading patients to be institutionalized were heart wrenching. Families struggled to make the best choice with hard decisions, the sense of abandonment when long term care facility was the only option left. There was still so much to learn, the treatments were never 100 percent effective.

After two years at Laguna and six more in private practice, Karen developed a theory that she was now intent on proving or disproving; finding another explanation for her years of observation of human mental functionality. Her breakthroughs were short when they happened and it seemed forever between them; but she continued on. In her initial premise, she knew there is energy in the brain. As with any energy, it must travel in certain pathways under the influence of certain rules. Planck, Maxwell, and Einstein had seen to defining the majority of these rules very well and the world has been only confirming them ever since. But her energy, beyond the standard wavelengths, seemed undetectable by any electroencephalograph or EEG machine and if it existed, it was not stored by any known storage methods. It was by all accounts, an impossible thing she was looking for, but her intuition overruled her training time and again while she continued on. Her theories had been modified over time of course, but her gut feeling was that there was something, just waiting to be found. Something of profound importance and usefulness to mankind, to be able to reduce or eliminate human mental suffering.

She returned to school to get an education in Quantum Physics and electronic engineering. She postulated that if this force, the soul or spirit, is an energy of some kind, then she should know as much about energy as possible. Perhaps combining her experience as a psychologist with a fundamental knowledge of energy and its characteristics, she could conceive how the energy she was looking for was being delivered and hidden from view. She believed even more passionately that it is there, just beyond the limit of her tools to measure. She could feel her closeness, but day after day she had been unable to define it, measure it, or describe it to herself. Let alone describing to the world who up till now had told her she was wasting her time. Leave the Soul to the spiritualists and Churches she was advised.

This was the advice of her mentor in residency at Laguna who had been her most ardent critic of deciding to take the journey she was now on. Dr. Lester Warwick was a man who recognized talent and used his influence to assure the students who passed under his purview were steered in what he felt maximized their potential for them and for society. He cajoled and argued when necessary, but his ultimate success in directing his charges would finally persuade even the most volatile student to at least try the direction Lester suggested. Lester had been in the business for over 50 years. Graduate of Harvard in clinical psychology, he had written over 20 dissertations and four books on the subject, three others on philosophical issues. He was the foremost authority on the human brain anomalies with conclusions that physical neurological damage is the primary cause of personality changes and dementia. Describe any injury to the head; Lester could outright diagnose the forthcoming personality changes, kinesthetic and motor losses, possible treatments, and likely prognosis. Karen absorbed his knowledge and found a respect for him she never believed she could have for anyone.

To Lester, Karen represented all he had dreamed of for helping in the curing of the mentally ill of our society. Karen displayed a tremendous objective empathy for all the patients, far more evident than any other student he had ever had, in 30 years of graduate teaching. Additionally she had an IQ of over 160, making her the rare flower in a million acre field he used to remark. When she opted to go out on her own, chasing God for no good reason as he saw it, he felt disappointed and let down. She tried to explain that her goal was not to define God. Her aspiration was to show the physical existence of some sort of explainable force or energy that described the physical behavior of people. In fact it was Lester's own influence which drove her to try. But as he was often known to do, he humphed and strode off, making Karen feel he was disappointed in her. It never made her feel good to hear Lester suggest she was wasting her time, but she also felt very passionate of her concepts and occasionally relied on JJ to support and comfort her.

JJ supported the concept of her going into private research and even helped write the grant application. He was so proud of her and her passion to find the truth and knew if given the time and support, he would have a Nobel Prize winner in his life. He supported all the efforts, changed diapers before Suzanne arrived, read her theories, was her loving devil's advocate, but most of all, respected her in her quest to change the practice of mental health. He had no idea most of the time what she was technically doing, but she could explain her complex work in simple ways, always engaging him by her actions, words, and excitement of her work. That was one of the fundamentals of their relationship, love and challenge, understanding and empathy. They were meant for each other.

2

This morning's celestial show and the warm glow she felt were not diminished when she downloaded the results of her latest overnight experiment. On first glance they once again indicated a negative, the energy she was hoping and expecting to see from her array of sensors she slept with apparently had not spiked in any way. She was ultimately measuring all frequencies of the energy spectrum for her unexplained energy. As she scrolled through her graphs comparing the common EEG frequencies, the Alpha, Beta, Delta, and Theta waves of normal fitful sleep with any other frequencies she could sense and record, they were once again fruitless, no energy increases or decreases. Her search was to find energy changes and correlate them with frequency changes in the very low brain wave ranges which were easy to detect and the source of known brain energy. This was at the heart of most of the criticism. Traditional therapists thought this was impossible or at least very speculative. She had been recording her own brainwaves for over ten years, and JJ's when he was in town, hoping to be able to detect her theorized energy or have a baseline when and if she found the changes she predicted.

As she was studying the latest data, the phone rang. It was expected, even though it was five in the morning, JJ calling from Zurich. He had been gone for four days now and always called to send his love to the kids and to Karen, while catching up with the daily events and reported status of his work.

"Well Good Morning sunshine," JJ said when she answered.

"Good afternoon to you, mien Herr," Said Karen.

"Any luck today?" JJ asked; he knew where she was sitting and what she was doing.

"No, so far I haven't decompiled the full run, but it looks like the run of the mill stuff. Normal patterns, nothing unexpected. How 'bout you; finding the new world order going smoothly?"

"Just the usual, Germany and France have convinced the European Union that they shouldn't pass this new ISO standard unless we include their new process. Only kings and queens of old could defend the obvious advantage that would give them. They may be an older established civilization, but they believe we are still children and try to get advantage over the rest of us. It totally goes against the Treaty of Rome and they admit it openly! That is the frustrating part, so now it's another round of balloting and probably another year before we will get this adopted."

"Well if your powers of persuasion couldn't win them over, there is obviously something wrong with their ideals. You can still whisper a mountain to collapse. How is your flight home looking? I saw the hurricane might make DC around the time of you connect."

"I called the desk; they say it's still looking good. I can't wait to come home and squeeze you professor. How are the boys?"

"Everybody's fine, except Stephan took to trying to fly, he swan dove off the swing and it looks like the black eye is going to be a real shiner. Happened yesterday when Jer taunted him, he convinced him to try after he saw a jet fly by. I think we need to have a talk to "Your" child about that."

"Our" child, he just got the name, not the behavior from me. I've got to run, the convener and I are going to have a talk with the German TAG, and this should be exciting. Keep at the research, you'll find it. I love you babe, see you soon." JJ always said he loved her, every time he had the chance.

"Love you too and come home soon, I need the squeeze!" Karen urged.

She hung up and began the second cup of coffee and pored over her results until the kids got up, which would be any minute now. She enjoyed her work, though it was like finding a needle in the solar system, but it seemed worthwhile and having a home life that allowed her to continue it and have a family was the best of all worlds. JJ had been left with a sizable inheritance after his parents died in a freak snowstorm just a few miles from their house, the house they now lived in. It afforded them a lifestyle that was free from the everyday hand to mouth living that seemed to haunt her childhood; never enough to get everything, but of course always enough to get by. Now they could get the essentials and live wherever they wanted and chose to live in Julian.

The Cuyamacas rose to over 6,500 feet and even though they were just east of mild San Diego, it snowed up there most years. In fact it is one of the local haunts for all of the Southern California natives to come to see snow; because of course you never see snow the flatlands. Julian is an old mining town that barely survived after the gold rush. Then silver and some copper were found and the town stayed on the map until they began to pan out. Ultimately the value sustaining the town was found by turning wooden sidewalks into tourist gold. It remains today as a historical reminder of life in the late 1800's. The streets bedecked in cobblestones and lined with shops behind wooden sidewalks, a quaint reminder of rural western life in the 1800's and early 1900's. Julian is far enough away from civilization to be remote, but still with internet cafes and a latte shop in town. Near the Cuyamaca Rancho State Park, with miles of hiking trails that change with the seasons.

The home they lived in was JJ's parents dream house. They bought the lot back in the 1950's and when they retired in the late 1980's, built the house. It was a two story wooden frame construction that matched the rustic surroundings. A porch surrounded the house on three sides. It was on a hillside, the backyard elevated above the ground with the front yard just below the street level. Its deep red clapboard walls and green shingle roof made the house blend into the woods. The yard was filled with plenty of pine and oak, set on a quarter acre of chaparral land.

The southern views overlooked the lake, with large

windows; the views were reminiscent of the pioneer days. Being on a hillside, the back of the house faced a gradually sloping pine forest. Trees so thick, the neighbors couldn't be seen. The second floor windows each had bird feeders which kept JJ's Mom busy with keeping up with the frequent visitors. They lived there until the accident.

They had taken a photo hike past Lake Cuyamaca, up the mountain for about ten miles. Around noon, the weather turned and it began to snow. The snow wasn't unexpected, but the amount was. In less than one hour, the trail was lost to two feet of accumulated snow. From the evidence, it was determined that JJ's mother had misstepped and fallen into one of the many craggy switchbacks that lined the trail. JJ's father followed her down. Both were found three days later by rangers assigned to the manhunt. It was a very sad day for JJ. He was the only child of the retired American Consul. His father was flown to Arlington for interment with honors. It had happened so unexpectedly. They hiked the trails there for years, it was so sad.

JJ had to set the affairs in order. Finding a life suddenly stripped of the closest people in his life was difficult. Karen and he had been married only a year when it happened. JJ took it very hard and she provided an emotional strength and the time he needed to accept it and move on. They moved into the house after two years and had now been there for almost eight years. Karen voiced the desire to have her own lab and research room that their condo did not have, so moving in was a mutually beneficial decision. JJ's job changed to one that consisted of more travel than not, his commute was only to San Diego's Lindbergh Field and back. Most of his work was by email and required little by way of an office; Karen eventually got the grant money and began to design her field experiments.

They settled into the house and fell in love with the nature which visited them on their doorstep. A far cry from the condo they started marriage in Lemon Grove. JJ was an international business major at UCSD. Karen was finishing her physics degree when they met during a psychology conference in Karen's line of work and which JJ needed to fill out his major. Karen, being the pragmatic of the two, says that when they met it was not love at first sight, but an instant pheromone attraction caused by the timing of their meeting and the state of his nearness to a sexual peak during a brief hiatus between finals. JJ says she was just plain hot and he wanted her from the first second he saw her. With the learned ambassadorial skill he was raised with, and his business acumen required for his major, he was able to talk her into going to a local coffee house and cajole her into a movie. Once the awkward first steps were taken, they both recognized the fit, the comfort, and the likelihood they would grow to put their teeth into the same jar every night when they got older. They were married the next year.

Their careers first took them both to different parts of the country. Traveling was a way of life. Karen insisted they spend their time early in their careers being successful, planning to have kids when they were more comfortable and stable. JJ had tremendous success. Everyone attributed his success to his tremendous analytical prowess. JJ always figured he practiced the sage advice of his father. He always recalled his father saying that diverse points of view were a constant and a seed of truth was found in them all. Look for the truths and gain the trust of adversaries. Practice patience with all people. He adored his parents. They were social, intelligent and very loving; to each other and to him.

JJ studied international business and quickly learned that cultural differences were the most difficult to overcome in a world where reductions to barriers to trade were the future. In 1959, the Treaty of Rome outlined the methodologies of removing the trade barriers which kept a financial segregation alive in the world; a true barrier to classes and limited job growth and equivalent value of goods. The beginning of the ISO. His knack for getting agreement on tenuous issues was an art, one that must be innate and practiced. It is almost like the ambassador role his father was so successful at. The business side of the equation was the one input JJ knew well and could use to get consensus, even where the differences had traditionally remained stagnant. He could clearly project the balance sheets for both sides and win over some of the tightest deadlocks.

The hardest issues to overcome were the cultural differences and their influences. Which side do you put the steering wheel on? How many people will fit into an elevator; religious differences which impact whether one side yields to another on issues. He enjoyed the pressure and the role he was often asked to fill, to help mediate the positions and come to an agreement. But no amount of training would overcome some of the most pigheaded people he met.

His upbringing and his marriage had given him a powerful tool of understanding the psyche of most people. His experience with different cultures was now the passion keeping him in his craft. The desire to be home with the kids and Karen was usually the detail ending the deal making. There is a point in all deals that all sides need to go home and dwell on the negotiation details. He loved his job, but he really did love his family more. He had often spoken to his father and believed that advice was the biggest influence in his success. He influenced a small part of the world trade machine, but it was a fascinating job and he loved it.

Karen would often ask him for his opinion on some issues; JJ used her in the same capacity. Karen and Suzanne loved the time after JJ's trips to listen to the personalities and cultural foibles he was forced to walk through. Laughter was truly the tone of many of JJ's tales of international travel. Though he wanted to be with his family, he loved his job and the time it cost to attend the meetings and conferences.

3

At the Social Disorder Institute in Bremen, Dr. Hans Schick began his routine of calibrating his EEG before use. Calibration of the device was mandatory of course, but being a type A personality made it a surety that the machine would perform flawlessly. His office and private practice was for the wealthy and exclusive clientele; socialized medicine was good for the masses but the best health came at the best price. Hans learned that if one wanted real wealth, one had to perform a service that the wealthy found impossible to get elsewhere or too tedious to go through normal channels to find. Hans found a niche and was very good at it.

Today's appointments included Mr. and Mrs. Lemoyne. The missus had recently had a baby and she was suffering what most would have diagnosed as Post Partum Depression Syndrome. The husband wanted a more thorough diagnosis. Her eccentricities were becoming worrisome, and he simply wanted the best answers money could buy. They were on time and the receptionist whisked them into the private waiting room while Hans prepared the EEG for his recordation.

Hans had been an honor student through university. His brilliance was only shadowed by his desire to improve his station in life, even if it meant climbing on and over his peers. He was accused of arrogance, of lauding his home school advantage because his father was a doctor. He railed that he was as good as or better than the professors. He did not change his answers when there were differences; he would argue that he had more insight into the subject than scholars on the subject, let alone the professorial staff, or his fellow students.

Some of his ideas bordered on quackery said one of his professors. "The brain is a wireway, but only within the borders of known physics. Only a madman would have us believe that there is some other unknown and as yet unpredicted energy at work. He is mad." Hans was looking to distinguish himself, to find somewhere no one had explored or discovered. His dream was to find something new. When he mentioned some of his ideas, his professors advised him to get some clinical time under his belt before trying to find an improbable solution. But Hans was impatient, for many reasons.

Hans finished Medical School and got an Electrical Engineering Degree and ultimately became an Electroencephalograph diagnostician for the Gesundheitsfürsorge. This afforded him practical experience with patients in the health care system until he ultimately opened his own practice. For years he had accumulated recordings of brainwaves in search of something. That something was Han's quest. What his professors never knew was that Han's grossvater had been in the Waffen-SS during Hitler's war, the glorious war.

The Waffen-SS was the combat arm of the Schutzstaffel; headed by Heinrich Himmler who was ranked Reichsführer-SS. They saw action throughout the Second World War. They eventually grew into a force over 950,000 men. At the Nuremberg Trials, the Waffen-SS was condemned as part of a criminal organization due to overwhelming participation in wartime atrocities. They enlisted many doctors for human experimentation because they had millions of human beings to practice on in concentration camps.

He had been a doctor doing research on Menschlich-Energie, human energy. The experiments were top secret and were never released, even after the war was over. Hans had learned of his grossvater's occupation at his deathbed when he confided his past. It was partly a confessional of his work at Bergen-Belsen, without asking for absolution, and mostly a plea for Han's to continue the important work he had started. Han's was 18 years old, looking for a career and as luck and fate had it, his grossvater had already decided for him. Hans swore to continue the work, secretly. He could not reveal the secret for fear that the wrong powers may acquire even the idea and control of the energy would be lost.

His grossvater described an immense opportunity and Hans became an instant zealot in the Aryan cause. If he could only find this energy, it would assure that the world leaders would be German; as it had always been meant to be. Like all the important discoveries made by the Germans over the centuries; and the fraudulent credit of truly German discoveries by foreigners; this one would change the fundamental way in which the world would operate. Hans would show the world the power of Germany and that true leaders and Aryans would be in their rightful place, and soon.

Mrs. Lemoyne was back for the fourth and final time. Today's EEG would be compared to the baseline established in the first two recordings. He had recommended Paroxetine, an anti-depressant medication, which she had dutifully taken. Today would confirm the dosage and correctness of the medication Hans had recommended. He applied the neural net while she discussed her recent life's happenings. Mrs. Lemoyne reported she was happier now. She kept her newborn with her at night, near her for her warmth. Her husband recommended it to allow her not to feel the separation anxiety she knew so well. The problem was her child had been two months premature and her doctors advised against sleeping together for fear of hurting the baby. So they fashioned a soft cradle which even if she rolled on top of, the baby could not be crushed. For the first time in two months, she felt content again.

Hans had given a two-week portable recorder to record and establish sleeping patterns and she was to keep a diary of the events and emotions of the day. This in combination with the EEG recordings here in his office would allow Hans to interpret the results as part of his ongoing diagnosis and ultimate prognosis the Lemoyne's were looking for. She described having her baby physically near her had changed her whole outlook.

She discounted any drugs working, though Hans knew it was likely that than anything else.

Hans noted all her comments mentally knowing that the automatic recorder installed in his rooms would be transcripted later by staff and waiting for him after patient hours, at 1500 in his office. He completed the last of the series and after discussion with the Lemoyne's, sent them on their way with the promise of a prognosis after he evaluated the portable recordings and the latest EEG results. He explained to them what he saw today was a good sign, her improved outlook and positive attitude. It appeared the drug treatments were effective.

The Lemoyne recordings were another bland set of nothing results; the force he sought was not detected yet, lamented Hans. His grossvater had been so sure, he advised him what to look for and yet after thirteen years, he still had no clue what he was looking for or why he had not yet seen anything approaching what his grossvater described. Perhaps it was only the rantings of an old and senile man after all thought Hans. That might be expected given the life he led.

Hans spent the next hour reading the recordings from the Lemoyne woman, saw nothing unusual, and then went to his daily habit of reading the journals. Then, he read about newer model EEG machines and the advances in neural technologies. He closed his office at 1630 and headed home, another day done.

4

Karen was roughly halfway through reading her nightly recordings when she heard the first bang of the soap dish hit the bathroom floor. "I guess Stephan is up," She thought. After he did that for the second time, she went acrylic with the bathroom toiletry. It took less time to clean up; though she thought it looked much tackier. As she turned the corner, Stephan yelled at the top of his lungs "Momma!" Racing barefoot toward her until he buckled her knee with the impact.

"Well good morning handsome!" Karen happily exclaimed. "I see you're up and ready for the world, where's Jer?"

"'Till 'tleeping," he yawned. "Did Daddy call for me?" he asked.

"He called and said he was coming home tonight to see you. I bet he has a big surprise for you!" Karen knew JJ was bringing home some Swiss chocolates and another stuffed toy that filled the hammock net above Stephan's bed. "Let's go wake Jer up and have breakfast. Today's a big day. We need all our energy today."

Stephan ran down the hall into their room and the sounds of jumping were only interrupted by the agonizing groans of an unhappy five-year-old being trampled into wakefulness.

"Daddy'th coming home! Coming home today!" Yelled Stephan into the unwaiting ear of Jeremy.

"Leave me alone. Get out of here," Cried Jer as Stephan pulled the covers off, just before Karen could stop the melee.

"Mom, tell Stephan to stop!"

"Stephan, don't wake your brother up like that, this is how

I want you to do it from now on." She sat on the edge of Jer's bed and gingerly kissed Jer on the cheek and said, "Wake up sleepyhead; it's time to rise and shine." She cooed.

Stephan just roared and said he would never kiss him, "That's groth..."

"C'mon you two, you need to show the new baby how sweet you both can be. Otherwise it'll be too scared to come out."

Both of the boys leaned into Karen's showing belly to listen for their new baby. Stephan began to talk to her belly, warning him not to listen to Jeremy; "Cauth he will make you do thingth to hurt yourthelf." His shiner still dark black and purple.

Karen interjected, "Stephan, you should know better than to think you can fly."

"But Mom, I thaw it on TV at Billy'th houth. Thuperman can fly and 'top a rock or a big tick. If he can do it, tho can I." He said.

"Well come on down to breakfast Superman, even he ate breakfast. And Jeremy, or are you Captain America, you come on down too. You'll need food for energy to keep up with your brother." She smiled.

Suzanne was already downstairs in the kitchen. She appreciated Karen's early mornings as she was the beneficiary of coffee already brewed.

"Good Morning you two, how did my favorite little brothers sleep?" She asked.

Both boys ran to the table where the warm waffles were waiting for butter and syrup. Karen said good morning and the day began.

Suzanne Vlavich was a visiting student from Dalj Croatia and signed up as an au pair with the International Au Pair Association when she was granted a student visa to continue her degree in electronics. She had opted to go to San Diego State and was lucky to have been selected. Karen met her in class and they were fast friends right away. Suzanne chose to leave Croatia after the deaths of her family in the Serbian War. As she explained the story of the horrible events leading to the bombing of her home, and her harrowing escape to Zagreb to live with relatives, Karen became totally sympathetic.

Karen had suggested they could use an au pair, especially when she is a degreed electrical engineer. It did not take much to convince JJ to consider the arrangement, with his traveling and Karen attempting research at home, having an au pair made the most sense of any option they had. Karen's parents were alive, but living in Normal, Illinois. Rather far away from San Diego where JJ was raised and absolutely loved living.

Suzanne had been with Karen and JJ for almost three years which was a relief to Karen and her mother who loved to travel, but not for long periods of time. Karen could not dream of returning to Normal after she had seen so much of the United States. She preferred the warm clear days of Southern California and she wanted to be with JJ, in his favorite part of the world.

Suzanne refilled Karen's cup of coffee and asked about the latest run. Above the noise and clanking of two eating boys, Karen said, "I have gone through the morning run and think the interferometer is either not calibrated finely enough or has a flaw. The scanned frequencies are off and on. They seem to be inconsistent."

"That's not possible. We retested them five times last night."

"But the results are, well, hold on; I'll get the printouts, be right back...," Karen said as she went upstairs for the reports to show Suzanne.

Suzanne was a very good electronics student. She had a minor in astronomy and major in radio frequency interferometry. This is the science of measuring the wave lengths of radio signals from different angles, enabling one to make much finer measurements of amplitude and phase shift by virtue of distance between the receiving transducers and the interference caused by the overlapping waves. Though discovered and used for light, further development revealed that much the same process can be used for radio frequencies. With this device, the further apart the input transducers are placed, the more fundamental interferences can be measured. Used on combination with an amplified EEG, energies both inside the brain and outside the brain can be recorded simultaneously in real time. Karen was working on just this concept when Suzanne came to San Diego for her continuing education.

It was Suzanne who designed and built the transducers and her and Karen wrote the software to make a detector for the hypothesized energy. Using the very high gain available and varying the recorded frequencies, Karen knew she would be able to measure it and once there, help patients by providing feedback or outright correction of the circuits which had gone awry. Karen held the highest hopes and one day she would make Lester proud, in spite of the bugger.

When Karen came back, Suzanne was picking up the plates from the table and pieces of waffle dropped on the floor as the boys tore off to explore the morning back yard; still in pajamas. First Karen laid the printouts on the table then completed the motherly duty of getting them in and their clothes changed. When she got back to the kitchen, Suzanne had already zeroed in on the anomaly Karen referred to.

"It occurred at 03:44:12; some sort of energy change, a dropout. But the interesting thing is that the drop is not total, it doesn't completely disappear but rather is only attenuated and then returns to the pre dropout level at 03:47:45, there," Karen pointed.

"I see; what freq were you monitoring?"

"Page 3, 2499.856MHz, we had moved up to the deep space bands last week." Karen fingered the frequency on the page as Suzanne turned back to the anomaly.

"But this dropout isn't associated with any solar events or satellite overpasses. I hate to think it's the hardware, but you may be right. I'll check it out after I clean up."

"It's interesting that this was the only dropout all night. I think it's worth recalibrating and repeating the experiment, though maybe we should do a broader band recording. I don't want to increase the number of sample frequencies too much. What if this isn't an anomaly and the duration of 3:33 is significant. Let's set up the sampling to five seconds for 45 freqs tonight around 2500MHz and see if it comes back. Meanwhile, I suggest checking the equipment."

5

As Geneva traffic goes, this was a breeze. JJ was going to be almost two hours ahead of departure time. That was good; at least there wouldn't be any needless rushing. He went through security and was on his way to an airline lounge when he stopped for the Toblerone and this time a stuffed, 150 mm penguin. He also found a child's Swiss watch with a white eiderdown wristband. Jer will like this he thought.

With his computer case, a shopping bag, and a coffee in hand, he casually walked to the airline club. There, to sit and open up the computer and work on the draft of his latest white paper. He sat with a thump as European chairs are lower than chairs in America. As he unloaded his laptop and retrieved his notepad to begin editing the paper, he took a minute for a deep breath, checked his watch, and took a sip of his coffee. He was grappling with the economic diversities countries have and their negative effects when entering into the European Union, or EU, were causing some economies to swing dangerously, uncontrollably, and unpredictably. He believed it was complex but manageable and trying to define it had been his mission for over two months.

The European Union is a federation of European nations which acts as a group to protect their economic and political interests in the global economy. With the larger countries already members, the smaller countries were struggling with the requirements of membership. Foremost was the change in currency, from the native currency to the EURO, in some cases changing the basis of the value entirely. If they did not complete the transition successfully, there were chances of oversight in value that the new requirements would later not allow full compensation for.

After an hour of outlining his paper, an old man sat next to him and struck up a conversation. He was very old, but his wrinkles had the evidence of being a happy person and his eyes shown with a gleam of knowing satisfaction. His clothes were crumpled and disheveled; they had the appearance of having been tailored for a younger version of the man. JJ concluded this was a man who wanted to age gracefully. His demeanor was no longer a concern to him; but it must have been once.

When he spoke, his voice was strong and did not fit the appearance of his age. At first he started off with a simple hello.

"Good Morning" He said, "Where are you off to?"

JJ was kind but wanted to continue working on his paper when the thought struck him he rarely took the time to be sociable. "Good morning to you too. I'm heading home. And you?"

"San Diego. I haven't been there for a long time. I'm meeting someone I haven't seen in a long time."

"San Diego. What a coincidence. That's my home. Were you in the Navy?"

"Oh no, no...I was Army, a medic; I've tried to be a peaceful person. Always left the fighting of wars to those who seem to need to do it. I'm more a, uh, philosopher of sorts you'd say today."

"A philosopher, as in a professor or as in a Sartre?"

"No, well, yes, I guess you could say both. I was a professor after years of having a psychiatric practice. Retired now, I was in the field for many years. Mostly clinical and EEG work. Since retiring, I only do speaking engagements. Now I consider the effects of mental health on a more philosophical level, the advantage of growing old. Perhaps no-one is aware that there are still positions which consider the direction of mankind."

"My wife is in that field. Were you a lecturer? Perhaps Karen has been to some symposiums you've attended." Said JJ.

"Maybe so, I have lectured. What is your wife's name?"

"She has actually left practice to do some research work, outside of the normal practice. Karen Jordan."

"The same Karen Jordan from San Francisco?"

"Yes, she worked at Laguna."

"Yes, I know Karen."

"Small world. What's your name? Jeremy Jordan."

"Pleasure to meet you. Lester Warwick. Karen may remember me; I still receive cards at holidays."

"Dr. Warwick, Karen speaks very highly of you. Of course she would remember you. So tell me, what have you uncovered philosophically about mankind lately?"

Lester thought of the question for a short time, deliberately considering the answer it seemed before he spoke. "I believe that as technology advances, the revelations may not be as welcome when we are able to distinguish truth from the myth."

JJ considered the implications before thinking aloud. "It is true, in business, politics and medicine, the more we know sometimes the more trouble we get into. So do you feel we should stop advancing, is that a realistic point of view?"

"I believe that all civilizations inevitably come to a technology boundary which historically has caused great pain or death for them. It's not for anyone to limit the advances; no-one could, only to prepare for the consequences of the discoveries."

"But then what is fate, if the predestined pain and death are all that we have to look forward to?"

"But we all have our final calling in the end do we not; it's only the how and the concerns of those who remain that are important. All life ends. How that life functions while living, ahh, that's the glory of life. There are those who have made fantastic tools only to have been killed by their use or masses of others destroyed have there not? What new culture has been created to allow the full truth to be discovered that puts not the discoverer at risk?'

"A culture that has more experience I suppose. Isn't it a matter of history repeating itself, with more experience the mistakes of the past become fewer and fewer...? You say this as if to know there is something afoot, something you know and are not telling. Anything I should know about?"

"Oh no, how could an old man know of such great things, I am only a theorist, I'm the voice, as you say, of history repeating. I manage my thoughts and try to be useful." Said Lester. "But consider hydrogen bombs, gun powdered projectiles, sharpened metal, wooden clubs tied to stones. The more you look back in history; each advance was the precursor of the end of their culture. The weapons which advanced one group over another resulted in a détente when all groups acquired the same level of weaponry. When one of the groups developed another greater weapon, the balance went away and the technology would eventually be transferred until the détente was again created. Such is the cycle is it not?"

Bringing the conversation to a sudden halt was an announcement of their flight preparing to board. JJ loaded his writing tablet, laptop, and notes into his briefcase. They both stood and began walking towards the boarding gate. Lester was silent and JJ reverted back to the business mode of the business of traveling. As they approached the gate, JJ was clearly going to board first and they separated. Lester said goodbye and he hoped to see him again and JJ concurred. As JJ handed the boarding pass to the attendant, he looked at the gate area, determined his best location to stand and Lester was gone. He must have sat down somewhere or went to the restroom JJ thought. With his briefcase and boarding stub in hand, he boarded and once again began to work on his paper, never seeing Lester again.

What a strange conversation JJ thought. In all his travels, most talk was regarding the last plane crash or mistreatment by the flight attendant, major sporting events, and the weather. It was refreshing to know deep conversation was still possible at an airport. He looked for Lester on the plane, just in his section, but didn't see him. He thought of the conclusion; cultures, détente, and advances that loom towards their own elimination. JJ wondered what new technology was on the horizon or if this were just the talk of an old fella who had taken the time to dream up a succinct way to state it.

Fourteen hours later, standing at the number two baggage carousel, he stopped to look for Lester, in hopes he could invite him to the house to see Karen and the kids, have dinner during his stay, but did not see him. "Well, I'm bushed," He thought. His bags finally came around the bend and he was off to the parking lot and the long drive home. He paid the parking attendant and headed up to the freeway. A big advantage of business travel is rarely being in the rush hour of a metropolis. The freeways were all at or beyond the speed limit as he raced east on Interstate 8, north on 67 to 78 and Julian, just twenty more miles at the 67/78 split. The trip was usually an hour and a half, traffic depending, today he flew. It looked like only an hour and twenty. He couldn't wait to be home.

6

Returning to his dining room, Hans finished his dinner with flourish. He was a gourmet chef, but only cooked for himself. Never married, he did not like the disordered lifestyle it created. His constant criticism of his partners had left them all with terrible feelings of anxiety until they would not even return his phone calls. Hans minded them ignoring him, he believed they should have the courtesy to return a call, but never once did he truly understand the original problem, that his hedonistic bullying and narcissistic view of life forced onto the people in his life was demeaning and insulting to them. So he opted for the lifestyle of paying for intimate company, company he controlled in every way. He found it more comfortable to be in charge of every facet of the relationships; and felt justified because of his psychiatric training. His belief that he was overly aware that others were not up to his personal rigorous standards and treated them as such.

His house was immaculate and featured a room dedicated to music. It was stocked with Wagner, Bach and other German composers' original scores he collected in specially designed cabinets. He owned a symphonic sound system that if he closed his eyes; he could imagine being there, with the composer, in a hall; without any of the mess that hundreds of people inevitably left behind. He often dreamed of being in the company of composers. He did not go to any concerts. Crowds were something he could do without; a disordered arrangement by his standards. He preferred to work, live alone, and dedicate his time to find this force his grossvater tried so desperately to relate to him before he died. His practice and his research were one in the same. With each EEG recording, he was looking in vain for some distinction, something his grossvater had described to him.

That conversation with his grossvater was an enigma, fractured bits of information; the most intriguing was his repeated claim of those who found this force would control the destiny of the world. He spoke about having found the energy in the radio spectrum, that there were ways to collect the energy and store more of it within a person, with the potential of making a race of Einstein's. Never again would the German race be subjected to anything but their rightful place as the leaders of the world. His grossvater was an obvious nationalist fanatic and though Hans had spent the better part of 13 years looking for this force, he never found any evidence of its existence. He also began to wonder if his grossvater had said they had found it or they had almost found it. Hans was not uneducated; he knew that if there was some force, it would likely have been discovered by now. After all, this is the 21st century.

But puzzles intrigued him, he went to school, tried to find any references to this force, waited until the government released thousands of documents of the Holocaust and he searched for any information of this research, grossvater spoke of. Once in his study, with his concertos at a reasonable level, he opened another box of copied documents he had borrowed from the Bundes-Bibliothek, the repository of all documents from WWII. With stout ale he began to retrieve folder after folder. He read of details of the purported crimes of many innocent victims, their only crime was not having been born Aryan. He found lists of family members who were all taken to Buchenwald, Bergen-Belsen, Dachau, and Auschwitz. He did not want to imagine their faces and with professional detachment, he could.

These records were three generations old now, too far away to affect him or consider the losses of those who suffered. He read dispassionately of the deaths of thousands of innocent people, guilty only by the heritage of their birth. As he skimmed the documents, these were copies of the originals; in the left lower gutter were the documents identification and sometimes a small note. He stopped immediately when after poring over thousands of pages, he saw his grossvater's name scribbled onto the page. It was the first time he had seen any scribbling on any page. This got his attention. He read the file from cover to cover, not just skimmed it. Olga Hoffman was the name on it but it was just like the last hundred in this box, similar to the last forty boxes; there were no other references or notes. He concluded this one person had been specifically assigned to his grossvater or that he had found something interesting in this case. He set it to the side and rummaged through the remainder of the box, about 60 files, when he gave up for the night.

7

JJ was turning in the driveway around 10:00 when he saw Karen's research lab turned off. It was late and that meant she wouldn't have seen his headlights when he turned into the driveway. He loved to surprise her by appearing in the house. As he picked up his suitcase he headed for the front door, it opened and there was Karen, arms outstretched, yawning, and pretending to sleepwalk into his arms. After a few greetings with no answer, he realized she may have been sleepwalking and began to gently guide her back to their room when she started giggling.

"Ha-ha got ya!"

"I can't believe I fell for that one. Baby, I got really worried. That was an Academy Award performance! Come with me and let me adore you all the way back to the bedroom." JJ said.

Karen raised her eyebrow and feigned an innocence "Sir, there will be no adoration in public here. I have a reputation to uphold."

"Madam, you offend, I will uphold every part of you, not just your reputation!" He reached behind her grasping the back of her thighs hoisting her up until she was a full head above his.

"Sir, if this is your idea of protecting my honor, I'm afraid it's far too late. You need to make love to me. Take me to our room!" She moaned in her Southern debutante lilt.

JJ chuckled; there was always a part of him that wished to be in her company every second of every day, believing every second would be just like the latest greeting, their role play acting. But even Karen knew it would not have kept a freshness;

a newness that neither of them wanted to lose if they both were stay-at-home 100% of the time people.

"I missed you," They said in unison.

"I'm glad I'm home," JJ exclaimed. "This trip was an ugly one, loads of infighting. I didn't understand Dieter; he has to be taking direction from someone. Normally he is more reasonable, but in this case he was much more vocal and outspoken against the proposal."

JJ downloaded his recollections of the interesting anecdotes about the people Karen only knew from dinner parties, until they were both undressed and ready for bed. In his pajamas, he went to the boy's room and tucked the penguin and eiderdown watch under their respective pillows, kissed them on the cheek and went back to his room. Karen and he lay down and massaged their day away; JJ stroking her hair and Karen scratching his back. It was a relationship built on a bliss and assurance of being together, one that led to making love every night they were together.

After JJ was in a love generated stupor, Karen smiled and set the instruments up for another night of recording. She and Suzanne had again calibrated the interferometer and tonight, there would be a repeat of the same frequencies and some others. They had eked out some more bandwidth by rewriting some of the basic code. As Karen felt virtually every night, like a fisherman who lays out the net at night, "Tonight I'm going to catch something, I just know it."

When Karen got up at 4:30 a.m. and started her ritual of making the coffee, stopping the recordings, setting up a backup of the data, running the FFT through the data searching for anything like a pattern. Fast Fourier Transformation is a complex mathematical calculation which can determine if a signal is random or structured. Given mass volumes of data, finding some regularity in a series of frequency peaks; like finding the regularity of the use of the word *and* in Shakespeare's works is impossible unless some method of data crunching is used. FFT makes this type of number crunching possible. Nowadays it is a function in most all math programs, a call function in Excel, a call in Turbo C++ making it available for public and private researchers.

Karen was pouring the second cup of coffee again when she pulled out the results, again more dropouts, nothing that stood out that said *Hey! Look here!* Like she wished it would. Just more dropouts of varying lengths, unexplained.

"This is getting very discouraging she thought. I'm sure Suzanne and I did a thorough system check. What the heck is this?"

Suzanne was eager to see this morning's result. She knocked lightly on the office door. Karen smiled and tossed the stapled sheaf of papers by Suzanne's chair. It was from these two chairs that the research recording was planned, designed, built, and operated. "Good morning," She whispered, "Anything good?"

"Nothing that appears to be that shining light I was hoping for. There are some more dropouts. I can't figure out why it's happening. These aren't like the ones last year when we had that solar cycle, full solar wind for a day."

"02:13:03 to 02:14:23 and another at 03:46:47 to 03:49:51. Not the same duration, so I guess we could rule out some man made interference, although I guess it could be something as simple as a garbage truck with a really crappy coil driving by the house."

"Is today garbage day?" Asked Karen.

"No, I was just kidding, I know you are hoping for a result, you must have overlooked the fact that very few things generate energies at these frequencies. We're talking synchrotrons, cyclotrons, quasars, and CBR. I wish we could explain it with some domestic reason...,"

Karen interrupted her. "What is CBR again?"

"Cosmic Background Radiation is a constant stream of very high frequency energy, usually protons, which is believed to have been created at the time of the Big Bang and continues to radiate from the point but gets bent by the ever changing shape of the universe.

"That's right, these freqs are only cosmic. What if they are truly dropouts?!"

"I don't understand."

Karen smiled "I know it couldn't have been a garbage truck. The obviousness of that hit me right away, but not the meaning of the information here. What if there is an energy that is there and not there for random periods of time and the equipment is fine. Would that mean something?"

Suzanne listened, looked again at the data, "I don't know Karen. This looks like the dropouts we measured yesterday. Once there, gone, returned and steady for hours. Why would some frequencies just turn off?"

"If we knew that, we'd be closer to the truth I think"

Suddenly there was a familiar bang of an acrylic soap dish on a tile floor. Off again to the world of parenting, Suzanne kissed Stephan as she went down to the kitchen to start breakfast. Stephan kissed her back affectionately, saw his mother, and raced into her legs.

"Ith Daddy home yet?" He asked.

Karen put her finger to her lips and said quietly in Stephan's ear, "Let's surprise him. He's sleeping. I'll get Jeremy and the two of you can wake him up and thank him for your presents!"

Very loudly Stephan exclaimed "Pre'thents! Where!?"

Karen escorted him back to the room and in the pale light, he suddenly saw a new friend peering out from under the pillow. A penguin! He looked up at his other stuffed friends and formally introduced them to his new pal, Penny Penguin.

All that introduction took a while as Karen sat next to Jeremy and stroked his hair until his eyes opened to a smile. "Daddy's home!" Jer said and raced to do his favorite thing in the whole world to do. Sneak under the covers of his sleeping father's bed and tickle him to life. Stephan yelled wait for me and off they raced. JJ never had a chance to wake up normally, he had gotten used to his sleep interruptions. He never used an alarm clock, between Karen's research and the kids playful wakening, JJ didn't need one, and as a bonus, this one never needed batteries.

Jumping out of his bed in mock terror, JJ returned the barrage of tickling as they ran for Karen's protection. Karen had decided to remain neutral in this fracas, lest her boys find out how very ticklish she really was. JJ agreed but always threatened to let on when feeling particularly playful. In addition, she would remind everyone that the new baby was fragile. The new day marched on, breakfast with the whole family, getting ready for games and projects orchestrated by Karen and Suzanne, participated in by JJ and the boys. JJ took them for a hike looking for a special beetle Jeremy asked him about at breakfast. Armed with his *Peterson's Field Guide to Insects,* out and off they went.

Karen would usually go with them on their backyard adventures, but today she wanted to examine her last evening's results and she had several errands to run. She had a strong intuition that they were important. As she scanned the results while her new Dell XPS whirred and sang the hard drive buzz shuffle, she wondered what the dropouts meant. She and Suzanne had set the neural nets and calibrated them for the nth time.

The structure of the net was very simple. The nets were tied to the interferometers in a matrix of ways. Under the bed was a screen material, the primary net. The net was linked to the interferometers with each node then amplified for data plotting. The interferometers were spaced in a radius of 50 meters around the house. By triangulation, specific frequencies and their associated energy fingerprints, including their intensities, were measured, recorded, and compared to the primary net. Those were compared to the EEG neural net that Karen slept with. The purpose was to determine what energies were normal. She used her own recordings until she figured out what to measure and how to measure it. She knew once she got that down, it was a matter of measuring the energies of lots of people to determine if mental illness could in fact be quantifiable and therefore reparable.

This triumvirate of measuring technologies could record energy changes down to 2.4 x 10 E—15. It could hear a pin drop on Alpha Centauri Suzanne was quite fond of saying. Karen's theory was if a force in the brain, the energy, could be measured, she expected it to emanate from the body at some frequency.

Her attempts up to now had taken years but she was feeling a new excitement as it appeared her experiments were doing something different. How she wished she could see what was going on. It was only a matter of time before she knew.

She developed the idea from actual practice, with the thought that energy has measurable qualities; therefore, it must be possible to record it or the effects of it and determine if there is some therapeutic use, once the results were known. She wrote her doctorate on the subject and received her first grant from the University of California, totaling over $300,000, along with several other grants for research purposes; but now it was running out and she may have to give her theory up, or go back and ask for more money. The technology to record at the scales she was demanding was very expensive, but the payoff would be incredible. Her paper focused on the upside potential, the potential to reduce mental illness diseases and the billions it cost every year, to eliminate the pain and suffering. She was never far from the hope of stopping human suffering, even if it meant having to ask for more grant money.

8

Berkeley in the fall is foggy in the morning, and occasionally the fog burns off the flatlands before 3 p.m., before the next round of fog rolls up the sidewalks again for the night. Hank Burns loved living there. It had everything in it he could love in a place given his experiences.

As though to differentiate itself, Berkeley in the winter can only be described by fog not burning off at all. Of course spring is different, the fog is fickle; it fog may or may not burn off and then there is summer. That's when the fog generally burns off by 10 a.m., to return again most nights by 9 p.m. Most of Hanks friends from the east coast teased him that the only distinctive change of the seasons in California is when the color of the cups at Starbucks change; red near Christmas time and white the rest of the year.

Hank was 38, with a degree in Eastern Religion from UC Extension. He was fascinated by Eastern culture, having spent some time there with his father who was a teaching professor. Hank had the opportunity to see both cultures, appreciate both, and note the differences. It led him to understand Buddhism and the purpose of self and the finding of the inner self as the sanctuary. Nowhere can you look for the answer to Who am I, Why am I here except inside. Know yourself and you will know all.

The Berkeley weather did not matter to Hank. Berkeley had the appeal for its unstructured attitudes and liberal points of view. The liberal political nature wasn't the appeal; it was the ability to practice faith without bias that kept Hank here. He spent many of his waking hours in meditation. His goal was to find the final layer of human consciousness, what he believed could only be achieved by deep meditation.

He worked as a waiter at the Berkeley Yacht Club. He figured if he was going to work, it might as well be for the most tips. He started out at the recommendation of a friend and once the managers completed their interviews and compared notes, one thing they all noted was that he was unobtrusive, not argumentative, not confrontational, and therefore would be a classic waiter to the several layers of snobs who demanded such blandness from the wait staff. Of course the wait staff were never privy to the discussions of the matter in more than a general way, more applicants were turned away for being a little too forward, a little too opinionated, and a little too sensitive to criticism.

Hank served as a waiter for five years, still not even close to the top of the seniority list of the wait staff, but now on the dependable list and not worried about losing his job. He was satisfied with the hours because he was usually at work or at the ashram, meditating. He jogged, ate only vegetarian food, and sought *The Way* of the Tao. He threw the *I Ching* daily, kept notebooks of information on his progress, and sketched his meditative insights as pictures. For him, waking up in the morning was followed by a meditative pose for an hour. Shower, dress, and go to the ashram for quiet time and conversation with the master. Following that, another hour in yoga poses and exercise. He had lately been seeing strange light, lots of light emanating from a human form. The light was too bright to make out details, but as he meditated, he delved into the light to see what it was showing him.

His father was still a Professor in China and they spoke often. Hank about life back in Berkeley and his meditation, his father about the politics and changes in Shanghai. Hank's mother died when he was four from a rare form of bone cancer. He did not recall much about her today, though he thought it was her guiding him in his meditations. His father told him his mother was a wonderful person; but there was no way she would be guiding him there. She did not meditate, she lived her life ferociously. He said it was as if she knew her life would be shorter than most and she wanted to make sure she had done all she could do. When she finally passed on after the extended illness, he sold their beloved home on Euclid Avenue in the Berkeley Hills and moved to China as a Professor of American Studies, so he could make the adjustment of grief an easier lifestyle. He brought Hank, kept him in the American group on campus, and exposed him to Chinese culture when he was eight-years-old.

By 12, Hank spoke Cantonese, understood two other dialects, and had traveled around the world five times. He completed his high school in China, because of his father's employer, he was able to go to any UC college he wanted to. He chose Berkeley and stayed when finding an ashram he felt comfortable with. He lived very simply in a two room apartment on Durant Street, east of Telegraph Avenue. His furnishings were a futon, a prayer mat, and tables for candles. His kitchen was a tad more appointed. He preferred most American flavors to the bland diet of the Chinese vegetarian. He enjoyed one book the most, "Eggplant in 1000 Dishes."

His primary form of transportation was his English racer. He rode it in the fog and rain to work and meditation. He kept fit and lean and believed he was living the ideal life of the Tao, humble, caring, and honest. His teacher was impressed with the dedication and patience of this Western boy and took favors with him. He asked Hank to explain and deduce his visions, as each person who has the images is different. He believed that the destinies of people are revealed in their visions during deep meditation and one should reflect on them in quiet repose for as long as it takes to understand the meaning.

Hank reflected on the lights he saw, emanations from what appeared to be people. The lights were vague and shapeless; too unspecific to provide meaning he thought. He went to his teacher to ask for his wisdom to help understand what the Tao was trying to make Hank understand. But every time, this teacher only said, it is only within you to know what it means. Walk with your Tao to know the meaning.

9

Lester Warwick rode on the Trolley for miles. Its red color and slow movement meant nothing to the old man. He wasn't on it for the aesthetics or the speed. It was a time waster, but he needed time to pass. He knew when he needed to be where he was going and managed his time accordingly. Just another hour now he thought. As the trolley came to the end of the line, Santee, and its new shopping center, he strolled out to the main street of the outdoor mall. Bed, Bath and Beyond, restaurants, Target; the place captured all levels of the market he mused, as he looked at the rising sun and believed that this day was going to be a good day. It certainly broke the monotony of the usual. He liked his jaunts although it was getting longer and longer times between outings. It spiced his days up. It was always for business more than for social interaction, but in this case it would be more social.

He wandered about, Home Depot, Costco, an Office Depot. Forty-five minutes to go. Starbucks, the Olive Garden, maybe a snack while I wait he thought. He wondered if his meeting at the airport with JJ would have caused Karen to reflect on what she was doing. Thirty minutes more before he would have the conversation; he noticed the sunlight began to hide the shadows. The mall was awakening with the people parking, the doors opening, one of the busiest places to be when the stores begin to open. He knew what was going to happen. He knew how the conversation would go and how it would end up; trying to end what was going to be a problem if left unattended. Only fifteen minutes to go.

Lester had dreamed up many scenarios and none of them

looked good. Left to unattended forces, earth shattering, culture ending times were certain. The forces were going to be very unbalanced, in a way that no détente could ever compensate for. He debated whether he underestimated the culture. Could they know and still survive? He shook his head again with the knowledge that absolute power corrupts so there was no controlling the power when it got out. He had to stop it, with any resource he could gather. He hoped he could pass this responsibility on.

The time had passed so quickly, but carrying the knowledge had taken its toll every day. He sometimes wondered how it would have been so different if he hadn't taken the burden on. He thought of another life, one filled with stability, children, and a home life. He returned to the moment once again, he had taken on the job, he didn't regret it except when he took the time to feel it; greater burdens are carried by other people he knew.

10

Karen and Suzanne were on the road to make the weekly shopping trip to get supplies and groceries. Many times it was the only reason Karen ever got off the mountain, except those rare trips to conferences or when she and JJ bought the occasional ticket to a play or concert downtown. Suzanne was the free spirit of the house; she could go anytime and spent quite a lot of time on campus both in personal and educational pursuits. As they were making the hour long trek to the store, they talked about the latest recordings.

"No, I don't think it could be an anomaly now" Said Suzanne. "I don't go for the garbage truck theory, these frequencies are just too high, remember, they are in the extreme cosmic range, only generated in the background of space."

"I know, but...but why are they dropping out. They radiate from space, can go through anything, and suddenly we measure a drop out, not once, not three times, but now for five days in a row. There is no obvious reason for it."

"OK, if not a garbage truck, then how about the ISS."

"The what?"

"The International Space Station. Maybe there's some experiment going on that stops the radiation; no, I take it back. It might pass over us once, but not five days in a row. Though it may be an experiment with some sort of shield or something, although I'm sure I would have read about something like that."

"Knowing that the equipment checks out, that we can record it every day and night, and then sometimes it drops out; what other conditions are happening?" Asked Karen.

Suzanne started listing the obvious, "Well first we have only seen them at night; although we don't record all day; we should start recording for comparison; it only just started. It started when we changed freqs; it is the same frequency all the time."

"It never happened with any other frequencies we recorded, did it, in the last four years. That could be significant."

"Maybe, it could be there is no specific period to the dropouts; it could be the period is cyclic and we just haven't gotten enough data, maybe."

"That's a good thought, something we can sink our teeth into," Said Karen.

They were getting close to the mall now, down Mission Gorge to Cuyamaca, past the Olive Garden on to Von's on the left. Karen turned right, and then turned into the large parking lot of retail heaven, made a command decision: first stop, Target, shoes for the boys, and a few items, and don't forget the birthday card for Carrie, Karen's sister in Walnut Creek. Suzanne was looking for the new Diana Krall CD and the soundtrack to Le Miserable's, both recommendations from her school friends.

They parked the car and as Karen passed a shadow of a streetlight walking to the entrance to Target, it occurred to her that a dropout could be from the source not sending the energy or that something was shading it. Something in her intuition kicked in, a strong intuition that she felt on many occasions. This one stopped her in her tracks, causing Suzanne to ask if she was OK. Karen returned to the now and saw Suzanne's worried face and smiled to let her know she was OK.

"Yea, I just had a strong dose of intuition, almost like an answer was being put in my path to find." Karen whispered.

Suzanne asked, "What was it?"

Pointing to the shadow, Karen said, "What is a shadow? It is the absence of radiation caused by something blocking the energy. We need to consider the range of the drop outs and whether something is shading the radiation or if the source is not transmitting."

"You think the energy is being absorbed by something before it hits the net?"

"I don't know what to think, but I know how we are going to find out, this just illuminated which way to design the test, to check for shading or for source drop outs. I think it makes sense to check for shading now." Karen was thinking of her next move, another net, a bigger net, a net away from the first one; see if the dropouts are similar on the second one. Isolate the matrix, how big is the shadow...Note to self, make sure you write this into the experiment book, it seemed so obvious, but she remembered to temper her enthusiasm with the fact it was just a hunch, not empirical, that will come.

"We'll map it out when we get home boss, right now I could use a vanilla frappachino and a lemon tart. Let's go shopping!" Suzanne said excitedly and took Karen by the elbow.

Three stores, one coffee shop and five shopping bags later, they were making their last trip to the car before a planned lunch at their favorite restaurant, Mimi's, when Karen looked up and saw someone she thought was an old friend walking toward her. An older man, he was definitely the man Karen thought he was. Yes, it was Lester Warwick. She hadn't seen him in over six years. As one of his favorite interns, Lester would always be the favorite resident of Karen's. He was so fatherly and scholarly, Karen always felt at home with him. Now, after all these years, here he was. What a coincidence, she thought.

"Lester?" Asked Karen as she walked over to the huddled man. At first he seemed not to hear, but as Karen moved closer, confirming to herself it was Lester, she exclaimed "Lester!" So loud he stopped in his tracks and turned to look at the approaching Karen.

Lester craned his wrinkled neck around to see who was calling his name and approaching him. He smiled after a minute, his recognition of her beginning to appear to be flooding back into his awareness. Karen took his hand and pulled him close to hug him when he creaked out "Karen! How are you? My, it's good to see you."

"Lester, how've you been? Please join me, us, for lunch! My gosh; it's good to see you after all this time!" Karen was awash with memories, the debates, the prognosticating, the lessons she learned from Lester were what drove her passion to improve her skills and her profession. Lester kept her always on the cutting edge of learning capacity. He sensed the ability and designed his lesson plan to keep maximum interest for maximum learning. He was an artist when it came to teaching, she longed to be in the same room with him, now to be with him after all these years, she was thrilled.

Lester was emotional too; he began to speak of the old days, the grand old days, back in San Francisco. He asked her about what she had done and what was she working on now, "You were always my cleverest girl, I always knew you would do great things." He repeated over and over.

Karen, over her original surprise and fluster, said, "Lester I would like you to meet Suzanne Vlavich, a brilliant young electronics student who is working with me on the latest project."

Lester turned to Suzanne and smiled; a warm friendly gaze that made Suzanne feel very much at home. She smiled back and reached for his outstretched hand. His hand was warm, almost hot in her hand. Karen was under his other arm still with her arm around him like a daughter would be. Suzanne had the warmest feelings for him, it made her pause to consider her comfort; she had never met anyone with such an unarming presence. She wished she had the chance Karen had, to have been around him in his day, to have been the cleverest girl for him. What an uncanny man this is, and all I have done is shake his hand.

Lester accepted the lunch invitation and the three of them sat in the waiting area for thirty minutes catching up on the last few years. Lester spoke, but it was obvious he was much more interested in what Karen was up to. He laughed when he learned she was "putting nets under her bed to catch God now," He chided her to not have gone on to other "uses of your brain that would really benefit man."

It was the same old Lester; Karen recalled toward the end of her internship and residency. He inspired her to do great things, but somehow he always left her the impression to pursue her passion of choice was the wrong direction. But he didn't do it in a way that came close to indicating she lost his respect. He could so disagree. Karen told him they had been developing the electronics for the longest time and segued into Suzanne and let Suzanne take it from there.

For her part, Suzanne felt as though she wanted him to know her story in full, to maybe bring this man into agreement on the merits and importance of this project. Suzanne told JJ and Karen it was the best thing that had ever happened to her; meeting Karen at school. She was a college kid with no particular passion for a career but a love of electronics that she had from an early age. She only looked forward, she thought, to working for some computer company and maybe having a life of engineering other people's ideas in a non-descript company somewhere. It did not truly appeal to her. When she met Karen and when the conversations turned to original work, to important work, in a family environment; well she jumped at the chance and never once regretted her decision.

She told Lester about Karen's boys and though Karen was just showing, Lester feigned not knowing about the third child on the way and congratulated her. As they talked through lunch, neither woman really noticed Lester had not spoken of his latest activities, and it wasn't until they finished their coffee that Karen finally demanded to know about Lester.

"I finally left Laguna and was appointed to be a manager of a small think tank. Not a big place, but one that was to my liking. I travel a lot, get to meet many of the clients, still dabble a little in the practice, but retired from the psychology side of the business. There isn't much to tell, it is a retirement job I guess you might say."

"But where is the office? Where do you live?"

"I did well in practice, I have several homes, and it seems I'm always traveling so I usually say I'm at home anywhere in the world." He replied.

"Why are you in Santee?" Suzanne asked.

He looked at her and Karen and asked, "Your research is bringing you to new insights, isn't it?"

They both rocked their heads in unison, "Yes," Karen said, "It's funny you ask this today. For the last week I've found an anomaly, although it probably is just an equipment malfunction."

Suzanne piped in "No, I don't think so. Lester, I think you would be very proud of what Karen is finding. I would value your opinion. Are you in town for long?'

Lester's eyes never left Karen's. Karen felt sadness descend into the conversation, she didn't know what or why, but Lester seemed concerned about something. She thought about the years of searching, finally an intuitive moment and Lester is there in Santee California, in a parking lot.

Karen asked, "Lester, do you have time to come to meet JJ and the kids?" She wanted to show them off, but Lester said he had only stopped for a short time and had to go. He asked for Karen's phone number and promised he would call. Karen asked for his and Lester said he didn't have a phone. He was always on the road and didn't want the trouble of a cell phone. Now Karen was very concerned, maybe it simply is a coincidence, meeting this way. It struck her odd he would not want to meet the family, but he was still a busy man and it wasn't really a planned meeting. They hadn't kept in touch, other than Karen's Christmas cards addressed to him at the hospital.

Suzanne felt the change in the tone and tried to brighten the mood, "Well if you can't come today, I for one want you to come on your next visit to San Diego. I believe Karen will make you very proud with her accomplishments." Suzanne repeated.

Lester agreed his highest regards for Karen were still there and now since this chance meeting, were revitalized. "I'm sorry I can't make it today, I'll be back and will call ahead of time to make arrangements." He said.

He walked toward the Red Trolley Station and promised with a wave he would be in touch. He smiled and kissed them both on the cheek with a word of how happy meeting them today had made "an old man feel." He said he would be back in town in a couple of weeks and would certainly be in touch then.

When lunch ended it was after 2 p.m., though they both felt they had spent all day with Lester. Karen told story after story of her times at Laguna Hospital, and her favorite stories were always with Lester in them.

"He seems like such a good man, he would have made a fine catch, I wonder if he ever married," Asked Suzanne.

"I don't think he mentioned being married. At first when I got to Laguna, I worried about the staff, I had heard stories of some doctors putting the moves on the new girls, but Lester never made me feel anything but totally safe."

Karen continued, "No, he wasn't married. I remember him giving a talk to spouses of some of the patients. I remember he could calm anyone; that was his special gift. He mentioned to the spouses that he could empathize even though he was not married. I would bet he would have been a catch in his day though."

They finished the last bits of shopping and headed home. It had been a sunny, warm San Diego type day. To have it all; great weather, warm sunshine, great shopping, well what else was there. They both made the drive back to Julian and after unpacking the groceries and other goodies, they went out to search for JJ and the boys.

11

Olga Hoffman; from Baden-Baden, Jewish quarter, 28-years-old, mother of three, four months pregnant. As Hans looked at the record before him, he tried to latch on to his grossvater's meaning. This woman must have been important enough for him to have personalized the file. It also put him at risk later when he had to explain in Nuremburg, at the war trials, why his name appeared at all on that document. Hans had read the transcripts, his grossvater had been a children's doctor, and he was called to many pregnant woman's side during the war to ease their suffering. He openly testified he was never for the mass killing and what he was aware of; he spoke out against until his life was threatened by the SS, though that was not really the case he confided to Hans. He told him in private that he had lied about his support of the extermination. He made Hans promise to work for the Aryan race, to reestablish them back into the rightful position as world leaders.

As Hans read on, he put his scientifically logical hat on and made notes of the Hoffman file. In noting some of the attributes and cross referencing with the history of his grossvater, there wasn't anything that stood out. He pulled the transcripts from Nuremberg and noted the medical nature of his visits; also knowing the true nature of the visits was to find the energy. He looked for some clue to why this woman, or was it her family or family ties that made her so intriguing.

Olga Hoffman was married to a jewelry maker, the son of a jeweler. They were married in 1937 and were picked up on Kristallnacht; on the night of November 9, when gangs of emboldened Nazi youth roamed through Jewish neighborhoods breaking windows of Jewish businesses and homes, burning and looting. Synagogues were destroyed, burned to the ground. 7,500 businesses were destroyed; 26,000 Jewish people were arrested and sent to concentration camps on that night alone. They were physically attacked; 91 died that night resisting the police and rioters. It was toward the end of a reign of terror that had really started in 1935 with the insults and discrimination on a nationally approved level. The Hoffman family was shipped to a concentration camp for Sonderbehandlung or Special Treatment, a metaphor for death. Han's reflected; a jeweler in a Jewish slum. His bias reminded him that they were treacherous and sneaky; the Jews. What were they hiding? Something had made grossvater pull Olga aside for his experimentation. What was it?

Again he recalled the final discussions with grossvater. He made constant reference to babies and pregnancy; what did it mean? "The energy is in the unborn," He had said, "What could it be?" Was the fact Olga was pregnant part of the riddle? He pored over more documents to find a correlation, but the records of his grossvater's work did not exist anymore, they had destroyed when the Allies marched into Germany. As he listened to the frenetic sounds of Wagner, he mused over the facts he had. He had been a doctor; he specialized in psychology after being a pediatrician for a number of years. He discovered some force or energy he believed could be tapped which could be used to help the race. How could a pregnant Jew contribute to it? He reflected for a while and went to bed. So many puzzles. What was the relationship?

12

Hank got to the restaurant on time to dress, get the specials of the day from the chef, and have their typical meeting in the back on the reservations, table assignments and getting into the garb of the dutiful waiter. He had a calm demeanor which management liked and a relatively good memory, but lately he appeared to have a lot on his mind. There were mostly older waiters at work and they had never spoken to Hank of his Buddhist leaning, though he once overheard a conversation regarding how strange it was to have a white person go so far East, "It's like he thinks he's Alan Watts or something...," Said one voice, the other responded "He is a good boy, hard worker, why do you care where he goes to practice faith?"

Certainly when he returned to the US, his Buddhist leanings were not mainstream. Perhaps that's why Berkeley fit him. With only like minded people in his life, he really never ventured outside his faith. Faith he learned and relied on when his mother died and he found himself growing up in China with only the monastic life of education and his father's love and support. Much more pragmatic than the cursory Christian faith. Hank believed Buddhism required more actual effort for substantial amounts of time; patience and practice, unlike the little practice that Western Churches required, the so called Sunday faith. He knew there must be equally devoted Christians as he was to Buddhism, but they were definitely the minority in his experience. The time and effort given to the practice of faith was more intense with Buddhists. There seemed to be a goal, to find the Tao, where with Christianity, the goal was to behave in a way to find the entrance to heaven; in very diverse ways given the different sects of Christianity.

His devotion to meditation was so satisfying and complete, he could enter semi-trance states at will, and his explorations into the lights he was seeing were weighing on him most of his waking days. He no longer worried over finding quiescence; he could find ever deepening levels of peace. With practice, he felt he would be as close to the godhead as ever, now that he was breaking new ground in his own pursuit of deeper and deeper meditation states. This was concrete and worthy of continued effort to find the end, the ultimate state of peace.

He ventured ever deeper, to continue the search for nothingness, to enter states so deep, it was possible to mistake someone for dead. This is the quest, to find the ultimate quiescence and peace through relaxation and concentration of nothing. It yields such refreshing peace; it is akin to runners enjoying an endorphin rush. It's like a clean drug, one that hooks you just the same as any other high. Hank was now entering new states of awareness, seeing new physical and consciousness states that it was thrilling to know he was moving further in; yet the new spectacles, the sporadic lights, were also a curiosity. There were times he tried to venture closer to them in deep states of meditation and they moved from him. He knew he must let them find him, be one with them before he could see what they were.

Megan Stonecipher was his best friend, also practicing meditation at the ashram, also a waitress at work; his first confidant when he returned to Berkeley. When time permitted, when there weren't customers and on their breaks, he and Megan would often sit on the rocky seawall and "take ten" their version of a quick meditation. The slight waves lapping over the rocks were instantly soothing and conducive to meditating. Megan had not practiced as long as Hank, yet she was a dedicated student of the Tao and sought the quiescence Hank seemed to ooze after years of practice. Their relationship was only about their experiences in the ashram and Megan often asking how he could go in so deep and could he help her to get there as well.

Hank often detailed his experience. Lately he was trying to unravel the lights, the streams that came into his inner view. Trying to understand the meaning of it and Megan could only guess what he was seeing. Hank tried drawing it, explaining it, but it simply was a description of light beams of an intensity he repeated over and over as "brilliant", though not a brightness like the sun through a cloud, but a piercing, razor thin beam with thousands of others in the same proximity, something he couldn't make out.

As the shift started, it was still slow, Hank and Megan began shining the silverware, and arranging the place settings on their tables and Megan asked, "Have you spoken to Master Chingkoan about the images?"

"Yes, I have, yesterday. He explained they were on my path, the way of the Tao. He said I can only find clarity of meaning by looking within."

"That's hard," Megan said, "If you haven't a clue, then where will you start to understand?"

"I think if I can bring myself deeper, I'll be able to see the whole image, and then, I will understand. Right now, I think I'm just seeing the leg of an elephant, or something much bigger than I take in from the vantage point I have found. But I'm convinced this is a course I'll see the end of; I'll find the end of this path and ultimate peace. I have longed for the deeper truth, I, like you, was impatient, but realized that only with time and practice can one find the way. It's what keeps me going every day."

"I wish I could get to that level, it seems I can't get to make a jump, only for seconds at a time. I lose it when I cannot control my distractions. I know it is just practice, but I long to be where you are."

"Megan, you have just started. Four years is not 25, I know if you continue, you will see, but loss of patience will only hamper the effort."

"You sound like a teacher, but you're right, I do practice and work on patience. But that's why I began, to find the Tao, to know peace. I'm here because I didn't know the way and I'm impatient, well you know..."

"You are a brilliant woman; it's only for you to take one small step, followed by another after that. Whether it's today or tomorrow, the knowledge that the steps are necessary is the beginning of enlightenment." Hank said in his calm supportive way.

They sat for their 20-minute break in silence, Megan to discipline her mind to stay devoid of distraction and Hank, to venture into the light, the stream of such blinding brightness; it could only be looked at peripherally. Lao-tse once said a man is blind when he looks into darkness and into light. One has to find what the light has or has not illuminated, to find the meaning by degrees of shading. Hank got to his familiar grounding, and waited for the light, but saw none. He had to discipline himself to feel no remorse or disappointment when it didn't come, he simply waited.

He imagined a blue plane, a sky blue empty backdrop upon whatever came to mind would be illuminated. He emptied his mind of all thoughts, all feelings until he sensed he was a viewer, a clean empty sponge which would be there waiting for input from what ever passed in his field of view. In a Zen state, the mere thought of being anything defeats the purpose, he did not think of a sponge, only the attributes of one, of a blankness of nothing upon which anything on the blue plane would juxtapose itself onto.

Though there were small streaks, they were far away, and while he knew if he wanted to see them closer, the act of his moving, in his mind was enough to change the state of consciousness that it would be futile. He knew when the lights are near; he will see them as they intend him to see.

Megan had been trying to awaken Hank for over a minute; she knew from experience that Hank was a deep and intensely focused practitioner, so she was patient. When Hank opened his eyes, Megan had a hand outstretched to help him up.

"Time to go back to work," She said, and followed with, "Any closer this time?"

Hank yawned and said "I saw them again, not as intense as I do at the Ashram, but definitely see them. Still cannot make out what they mean, if that's your question."

She laughed and said, "I know if you continue, you will see, but loss of patience now will only hamper the effort," Repeating his advice to him with her smiling eyes.

13

Hans left his house, promptly, getting onto the train exactly at 8:34, got off exactly at 8:59, walked exactly 104 meters to his office. He knew the exact distance because he counted the steps at least three times a week; such was his nature. He did not waiver, and had mentally counted the number of times the stoplights favored him which he used to calculate his odds for the next day. He railed at the trains, if they ran on time, as they should; the stoplights should be the same every day.

Arriving in his office, sliding into his chair, he placed his attaché on the rear desk, turned the five number wheels to his unlocking code, opened it slowly after looking over his shoulder to the empty office, and closed door. He changed the unlocking code five times a month, randomly with no repeating patterns. In his experience, the importance of his Aryan leanings and the research into his grossvater's work were important to keep secret and he was very careful to rationalize this somewhat anal behavior as his maintaining his doctor/patient secrecy.

His first appointment was not until after noon, he decided to lock his office and visit the Bundes-Bibliothek ten blocks away. He was ready for another box of the historical documents. In his visits, his relationship with the old librarian was nil, he never spoke to him, and this disinterest was finally challenged by the librarian. Dieter Negel had seen Hans come and go with thousands of documents, box after box, for years. He never asked Hans what the purpose of the research was, he assumed he was doing a history book or a novel and needed historical references, but after seven years, it was obvious that the customary reasons were not typical. He decided to stop Hans and ask, "Herr Schick, I see you are back again. It seems you are looking for something; perhaps if I can be of help to you, your search will shorten, and you will have what you seek."

Hans was startled, "What is your interest in what I seek?" He replied sharply, as if to rebuff the man for being nosey.

"But I only am curious why all these years you search, I can only assume that you are yet incomplete if you continue for so long."

"I do not wish to discuss it. Mind your place. I have done nothing that merits your attention. If I require assistance, I will inquire."

Negel was astonished by the reply, in his years as a librarian it was his custom to offer help and be responded to positively; this man seemed far from his norm. It was not beyond his notice that the subject matter he searched through was the darkest of German history. It occurred to him that anyone who revisits such a specific time in German history, for as many years as this Schick had, with no other apparent interests, may have had a loose screw, a fanatic.

"Herr Schick, I am familiar with my place and was performing my duty to offer help, nothing more. Since I cannot be of assistance here, I will indeed move on. Good day sir," He said, he thought of sneering it out at this banal person, but the thought of Hans being a possible fanatic kept him as neutral in tone as possible. The newspapers were full of stories of Nazis and skinheads who randomly killed; he did not want to ire a person who his gut was telling him was not normal.

Hans turned his back and walked back to his section, the section of the library seldom visited by anyone. He did not spend much time there however, an hour here, two hours there; he preferred to get his box filled and leave. Fortunately these copies were not in the reference section, which would require he sit here, in the library, where good music and a lively cigar would be unavailable to him.

Wait, Hans thought, the reference section, I wonder if that original Hoffman file is here, perhaps there is more in the original file than in the copy. It meant seeing the librarian, but he is, after all, "just a functionary", probably not even a German he thought. He loaded his box with a new batch of documents and went directly to the reference area, where older original documents must be checked out only in the reference area, not to bring home.

14

Lester knew what he must do. He had thought for many years about what he must do, but could the subject be relied on to follow his protocol and keep the secret, a secret so important, he had in fact never revealed before to anyone. He guarded it by never having written it down; except in cipher and encoded, and by finding those researchers close to it and stopping its discovery by discouraging continued exploration. He had used many methods, but in the end he was successful. It had not been revealed. He alone had determined it was best to not allow the truth out for many reasons.

Lester had spent his long life as a clinical psychologist. He had been steered into the medical profession in the same manner as most in the field; their fathers had pushed like hell to keep the godlike function in the family. Lester was not to be a surgeon though, his personality leant itself to spiritual beliefs and healing the mind. It was cemented as a result of his experience in the Big War, WWII, when he entered Bergen-Belsen attached to the Allied 21st Army Group, a combined British-Canadian unit on April 15, 1945. At the time of liberation, the camp had been without food or water for three to five days. It was the most horrific sight he had ever seen. For the 23-year-old man, it was the essence of hell. No history book could have warned any of them of the horrors they would find. No preparation could have eased the vomitus blow of opening the doors to the barracks, that were in fact the ovens and holding pens for thousands of people, stripped of their lives as though they were ants. No regard for their value, their souls, or their young.

Lester was a medic; he got the first calls to see to a group of women in Holding Cell 43. A leaky, wooden, uninsulated structure where 114 naked and bone skinny women lived only by huddling close enough together that those inside the huddle would live, and outside were picked off by their captors for Special Treatment or frozen to death in the winters. But more than that, they were all pregnant. The miscarriages were pushed to the side of the cell. It was an image he could not get out of his mind from that moment on. He instructed the troopers to bring them all into the medical hospital they had set up outside the prison, with real beds, warmth, and life.

Lester began to heal the cuts, the scrapes, but he could not heal the blackness that was evident in their vacant eyes. Their flesh was again turning pink with the fire of life, but their eyes spoke of horrors and seemed they would never come around. Lester went on leave, home for a three month furlough, and met a psychologist friend of his father's. Lester explained the horror and asked if there were anything a medic, or for that matter, anyone could do for them. The doctor inspired him to become what Lester ultimately became, a renowned mind clinician, a passionate advocate and someone given a chance to help a group of people who seemed as dead inside as one could be driven to. He offered to volunteer for this specific duty and encouraged twelve psychologists to travel to Germany and Poland, both to help and to learn how to treat these terribly unfortunate victims of the Nazi's. Lester documented the experience and decades later added his expert input to the *Diagnostic and Statistical Manual of Mental Disorders*, the diagnostic handbook for mental conditions first published in 1994.

Lester went to the Division Commander and explained what he had seen and what he wanted to do, to bring the best doctors from the States he could and return to college and become one himself. After much red tape, Lester led a contingent of medical doctors to Bergen-Belsen and later the others; Buchenwald, Dachau, Auschwitz; to do what could be done. It was at Bergen-Belsen that he saw the worst of all inhumanities and where with the help from German doctors and translators, the search for

Mengele and his band of animals began. Many of the Nazi's escaped, but not the majority.

It turned out to be the most rewarding experience of his life, but not without an acquisition of utter detestation of the people who did these things. He determined he would be one of those who brought the unjust to task; he documented all the cases and was one of the foremost witnesses in Nuremburg. When that was completed, he completed his degree and began a practice which lasted his lifetime. His unique experience made him both seasoned at an early age and respected for the experience and his handling of it and the patients who eventually moved to the United States to leave the horrible memories behind, to start new.

He accepted a job at the Laguna Hospital in San Francisco where he later proctored a residency program. He spent two days teaching, one day of practice and the rest simply as an advisor to the staff. He had his favorite students; over the years he had followed only one after they left his care, Karen Jordon. He knew if there was the potential for anyone to discover the dark, powerful secret, she was able and had the resources to do it. He had to steer her off the track.

15

Suzanne and Karen brought in the shopping bags, called out for the boys, and found the house empty. JJ had written a note saying he and the boys "would be at the Cuyamaca Restaurant for dinner, join the food fight if you dare!" JJ knew Karen needed to focus on the hot issue of the dropouts. So the note went on to say, "Take all the time you need today. I want to catch up with the boys, love, me."

Suzanne commented to Karen that she hoped to find as nice a guy as JJ. With only a deep friendship innocence, Karen understood and in their time at school, Karen often became a sounding board for Suzanne's boy suitors. Most all of them failed right off the bat, the better choices had some lasting value, but in the end, Suzanne had yet to meet the right man and Karen and she spoke often of the time left to find the right one. Their mutual respect and trust was so rock solid they both relied on each other both personally and professionally.

Suzanne was already doing a diagnostic when Karen got to her office to add the results to a spreadsheet she could do a comparison of data with. Her goal was to find any correlation to the dropouts. As she was creating the worksheet to chart the data, Suzanne came into the office and asked Karen to wear the head net and lie on the bed to get a baseline of energy for later reference.

The entire system consisted of the interferometer with four nodes, 100 meters square around the house, at the center was Karen's bed with a net under the bed 2 meter square, under a throw rug. With the ability of the interferometer to amplify and record very small energies, it was possible to record any energy emitted from the brain and correlate them to internal brain energies. It was Karen's thesis if the brain could receive energies with normal wave characteristics, one could prevent outbreaks of psychosis and by wearing a cranial net or embedded electrodes, to control and eliminate mental illness.

Her theory had been challenged after the years of research she invested and now that there were some anomalies, it seemed the needle in the haystack was finally coming to light. Suzanne ran through her diagnostic and found a base energy level, confirming the frequencies were the same and instructed Karen to position herself over the floor net. As Karen sat on the bed and began adjusting the cranial net, Suzanne called out in surprise.

"What did you do? The readings just dropped out!"

"Nothing, I sat down."

"Look at this reading," Suzanne said.

Karen rushed into the office and saw the same drop out pattern she had seen from the overnight recording. "What caused it?" Asked Karen.

"I don't know, but it stopped when you came in here. Go back onto the net," Suzanne said.

Karen sat back on the bed and called out for Suzanne to push the telephone intercom button. "Did the drop out return?"

"Yes. Step on and off the net Karen!"

Karen stood and took two steps which took her off the net, counted to three, stepped back onto the net, counted to three, and repeated it several times. "I'm cycling on and off the net, are you seeing a change?"

Suzanne said, "It looks like you have several three second cycles; I assume you were stepping on and off, it looks like you did it five times. I have to think that the dropouts are from your movement, but how can that make sense? And further, it happened without the cranial net, so it wasn't due to any wave correlation from our equipment. I'm stumped."

"I'm going to the other side of the bed, perhaps the net is damaged on this side."

Karen repeated the on and off counting to four instead of three and after several cycles asked Suzanne to report.

"It looks again like you are cycling on and off but the period is now four seconds. Karen, it is definitely the influence of you over the net. Let's trade places."

Suzanne went to the bedroom, spoke to Karen, and began cycling on and off the net. Karen reported there were no dropouts.

"We are talking about cosmic rays being stopped by something about you. This can't be happening. If the other dropout periods are any indication, then we only have minutes to try to figure out what is causing them. Go back on the bed and strip down. Let's see if I can pinpoint the location on the net where the dropouts are occurring. I have to remove the averaging algorithm, go, go! We can be recording this while I'm futzing around!"

Karen had her clothes off, put the intercom on speaker and began a three second cycle for reference then climbed into bed. "I am on the northeast corner of the bed, standing, I am walking, I am now on the northwest corner, I am now on the southwest..."

"Lie down in the middle of the bed!" Suzanne said.

Karen lay as close to the middle of the bed as she could be. She had her arms and legs spread and used those to center herself. "OK, I think I'm centered, can you tell where the dropout is happening?"

"Shit, I had to reboot the interferometer program to remove the averaging function, it'll take a minute. I still have a definite drop out going on with the CR recorder; hold on, its coming online, I need to recompile...it's gone."

"The drop out?"

"Yea, it's gone. What can it mean, what is changing?"

"How much did we get recorded?"

"Up till I had to turn off the IF amplifier to try to get a quadrant fix. I never thought I would need to know the location of the energy on the net, only the levels of radiated energy. I will be ready next time though."

"What do you think it means?" Asked Karen, to herself, but Suzanne answered, "I know that cosmic particles can go through anything except heavy water a mile deep. I know that you're not built of deuterium, so therefore there is something else that stops CR and we know nothing about. But the weird part is that it was something in you and not me. I'm jazzed that we have something to bite into now, there are so many things we can do now that we know what we are looking for, harmonics, embedded frequencies, nodal analysis, and comparative string values analysis. I hope I recorded enough data to begin some analysis. By the way, that was clever thinking, cycling on and off. I would have to say it empirically proves the change was caused by your location, presence, over the net."

"What does it mean?" Karen asked out loud. Was this normal; was this dangerous to her and her baby? What is it? It would take quite a while to write out the experience in the notebooks and it was such a find, it had to be repeated. Her electronics training did not prepare her for elemental physics analysis; fortunately Suzanne was a great resource. But first, write down the experiment and write down the results; first rule of science; document. Conjecture later, just write the facts for now.

Karen went to her computer and related the events; Suzanne went to her equipment and tried to coax answers to questions neither knew what to ask.

16

JJ and the boys were fishing on the lake, mostly catching sunfish and no real hope of anything for dinner. The seasonal Lake Cuyamaca had recently dried up when drought conditions lasted for more than five years. The dramatic weather change had long-lasting effects on the whole State of California, even political.

When the searing summers wear on and little or no relief from the heat is in sight, fuel prices had a fabled biorhythmic triple low and as the State was experiencing brown outs, so too was the Governor browning out, unable to remain elected. Weather always plays a part in the San Diego area, water is a commodity and precious. But always a yin/yang relationship, when the rain comes, the vegetation grows and becomes fuel for the next wildfire. The Cuyamacas were a testament to wild fires.

In the last three years, Julian had gone through two calamitous fires. Volunteer fire departments from all over the western United States assisted in keeping the town of Julian alive, even at the cost of one fireman. To say the weather does not have a role in life in San Diego is to not understand the nature of cause and effect to its elemental values. Lake Cuyamaca is just one of the natural lakes which can illustrate the values of nature.

For JJ and the boys, it is now a full lake after the pineapple express spent two months raining on the San Diego area. Beautiful to see when it is full, feeding the animals and rushes, causing a frog cacophony at night, also a body teeming with sunfish and stocked bass sportsman hope will grow into the trophies while the lake level lasts.

Stephan had his hook and bobber out for only seconds when he again yelled "I gottem, I gottem! I got more than you!"

Jeremy had tallied differently and said, "No you don't, I got ten and that is your ninth!"

"Ith not, Daddy, tell Jeremy I got more than him..."

JJ could hardly keep up with the count himself, he was either rebaiting a hook or unhooking and releasing. He had not even had a cast of his own. The early afternoon seemed to bring all the hungry fish to his craggy nook of the lake.

"I think it's a tie," Said JJ the mediator, and they both complained that it wasn't right. Just as Jeremy's line went taut and buried his bobber.

"I have never seen the fish like this before, even when my dad brought me here."

"It's because of the stocking program" Said Karen.

"Mama!" Screamed both boys "I got more fith than Jer," Screamed Stephan louder then Jeremy.

"No, you didn't! I had eleven and you have ten! Mama, he is always saying things that aren't true!"

"Do not."

"Do to!"

"Do not!!!"

JJ broke into the match and asked why Karen was down at the boys secret spot and he winked at Jeremy. JJ always assured the boys this was a boy's only zone whenever the girls were being a pain. He meant it to be a special place where he and the boys could always go anytime. Jeremy would even tell Suzanne and Karen he was going somewhere he couldn't say because only the boys know it.

Karen had spent three hours detailing and documenting the experience and had to get out of the office to reflect and as she was hiking she heard the boys talking and decided to stop in and watch, but also to talk to JJ when the time was free to have some adult conversation.

It wasn't unusual for Karen to hike out and find JJ and the

boys, but there was something in her eyes that made JJ aware that something was really on her mind in a serious way. JJ told the boys, "We should go over to the other hole, if they're biting here, they will jump into your open hand in the other spot!" It was really a distraction to have a few minutes to talk to Karen. Stephan was reeling in as fast as he could to beat Jeremy when they all stood to go.

"Beat you," Reeling in his line.

"No, you didn't."

"Yeth I did."

As they competed in everything, it was easier to let them hash it out themselves which JJ and Karen did in a mutual parenting agreement. Walking side by side, JJ asked Karen "What's up, doctor?"

"I had a really unique and weird experience and I'm not sure what it means."

"What happened?"

As Karen explained in the staccato way parents do with children constantly interrupting, JJ first thought enough to try to alleviate Karen in her worry of this hurting her or the baby. "Simply because you have measured something doesn't mean it wasn't there before. It has to be a natural phenomenon."

"But most people don't wire their house with neural nets and wear cranial socks; I'm concerned there might be some secondary effect going on, something cumulative..."

"Stephan was born with all the equipment and he looks to be normal." This caused them both to look at their youngest as he was about to eat a caterpillar.

"Stephen! Do not eat that!" They said in unison.

"Well OK, but that is still normal," JJ laughed as he chased the boys and put Stephan on his shoulders.

"For the first time in my life, I'm worried about hurting our children. I never gave it a second thought before now and I feel that I'm doing something to the kids, I'm scared."

"Baby, from all I know about your research, I can't believe there is anything to worry about. You're measuring rays of energy and when you stepped on the net, they stopped. That says to me that those rays were there when you were off the net as well, and the IF equipment didn't steer them to you. They were there all your life; I don't see any reason to worry."

"I know you're right, there is just a nagging feeling that there is something to worry about. I don't know...I've thought about the timing, stepping on and off, I don't really worry because the nets are only passive, the IF field is new, but it's again only passive. My gut is also telling tells me it's an important breakthrough, but I haven't grasped what it means. I need more information."

"Suzanne is working on it right? Let it go for now, she can think on it a while. You and she are a good team. Relax and enjoy the day. Suzanne should be here as well, but I guess after what's happened, there's no way to get her away from the computer."

"Thuthan thould be here with uth," Said Stephan with his perennial lisp.

"She'll come when she's ready to," Karen said to Stephan, "And not until she has some kind of answer. Let's go to your famous spot, where the fish jump into your lap and then go to dinner." Karen said, but absentmindedly, she was still very much in thought JJ noticed.

17

Lester knew that Karen was close, he expected her to have already determined the energy was a fundamental in the universe, especially since she had the latest instruments to test with. Lester only had an idea of what the truth was; he didn't know the how's, it was the why's that worried him the most.

Back after the war, Lester was assigned to Nuremburg duty and had access to all the documents confiscated from the concentration camps. Documents that proved the existence of the genocidal intent of the Nazis, leaving many examples of how to perform the brutal acts on paper. There were many papers that were dreamland stuff, so far out that anyone reading them would have only guessed they could not be true or possible. It did indicate the depths the Nazis were willing to go to, to eliminate everyone but their own race.

Mengele's human experiments, Himmler's torture experiments, and the one thing that he hoped would never come to light. The secret that would turn the world upside down. Genocide on a total scale and the bonus of taking life itself from one group and adding that force to another. Some of the documents had details of concepts that were unavailable to prove technologically, but that was changing now.

18

At the Institute of Human Study at the University of California Berkeley, Dr. Roger Hadlyn was finishing the planning of this years Forum on Human Studies. The lineup was excellent, he had managed to convince one Nobel Laureate, five tenured professors and ten leading researchers. This year, the focus was on brain research with a special emphasis on EEG advances and the latest advances in neural enervation in the brain. This was one of the most hopeful fields of brain research that included many non-drug, non-surgical therapies for prevention and curative brain diseases in brain wave function.

Among the keynote speakers was Dr. Karen Jordan, whose individual research had already provided insight into the cause of theta wave variation causing several dysfunctional sleep behaviors. Her research was in several leading journals and most important for his forum, she was so approachable that she was always asked for by the Forum participants.

Today he was finishing the Forum publication for mailing out to the registrants. As he proofread the bios and abstracts, he was excited when he read Karen's. She was going to detail the results of her attempts to use a neural net to record brain wave patterns for use in diagnosis and begin preliminary experimentation in inputting magnetic impulses at specific frequencies to control unstable wave patterns. Her dedication to this line of research was well known in the field and her results were always the talk of the forum.

He hit the enter button of the computer to send the files to the printer and uploaded the website with the links. He went to his office where he saw the message light blinking and saw four messages. The first two were from his daughter reminding him of the dinner tonight with her family, the third from a grad student giving a lecture next week wanting advice on a topic and finally a call from a German man asking if his Forum confirmation had been processed properly as he had not yet received word. Roger laughed as he thought of the German intellect of exactness and formality, this person, Dr. Hans Schick, had only called yesterday to reserve a slot. Not that it took a day to sign him up and charge his Visa, but his call was a little quick in the scheme of things.

He emailed Dr. Schick that his slot was in fact secured and called his grad student and moved on with the day. There are still details that need to be followed up with, the dinners, sound, and video production; these Forums are always a lot of work. Even with help from students, the details were endless and he wanted to not have to rush around, last minute, making something work that was forgotten. This was his fifth forum he was chairing; he thought it would go smoothly this time, enough experience now to know to get all the detailed work completed beforehand.

19

Hank and Megan were out riding their bikes through North Berkeley. Their favorite ride was up Euclid Avenue to the Rose Garden, stop for a picnic, and finish with a slow downhill coast back to their apartments on Durant. On sunny days, the bay would sparkle with the slot winds causing afternoon waves to crest. The slot is the area of the Bay that acts as a wind tunnel for the high speed ocean air as it is pulled into the Imperial Valley due to the heating of the valley. They could approach 60 miles per hour and are a sailors dream. Dotting the bay were sailboats of all kinds. In the morning, the sea fog fills San Francisco Bay and through to Martinez, through the Carquinez straits, on foggy days, it was a sight to look over the fog and see Coit Tower on Russian Hill, the high rise buildings and the Golden Gate towers above the cloud deck.

Megan met Hank at the ashram and later they were surprised to learn that they lived within a block of one another. They shared many common beliefs and politics, their spiritual paths crossed as well. Megan was amazed at the focus Hank commanded both in meditation and in real life. He was fluent in Chinese and was a selfless, considerate person who she was attracted to from the day they met. She enjoyed his wit and even on the rare occasion he displayed it, his charm. He was so matter of fact; it was often mistaken as being harsh and dry.

But Megan knew better. She had seen him when he spoke of his mother, of the pain when she was gone and how his father became his life. She could hear the echoes of Lao-Tse when he related his monastic experiences in China. It filled him with a viewpoint that at first seems irreconcilable with western living, and he was quiet in his adjustment. Megan became his transition advisor. She looked to have what he had learned in his youth, and at first he looked to her to help adjust to the complexities of modern life in the US.

They decided to stop for a pizza on Ridge Road, just north of the campus. A local student hangout, they were not students, but didn't look too old to look like they could have been.

"What do you feel like? Veggie or cheese?" Megan asked.

"I could go for cheese today. Why is it so empty in here?" Hank wondered.

"Maybe it's the football game, they're playing Stanford."

"Oh yeah, dahhh. Well, we get the good seats today, by the window."

"I got it." Said Megan when the pizza order went in. Hank smiled appreciatively.

"My treat next time," As he carried the drinks over to the table.

The window was covered with posts, some for sale ads, some advertising for various services, and tutoring. Others were some of the University events with dates and times. Megan had gone to the bathroom and Hank was randomly looking through the newspapers and flyers. One piqued his curiosity. A forum on the brain; Center for Human Study; Brain Function, Latest Developments. Swami Shaolin Koan was one of the guest speakers. He had spoken at the ashram, now he was speaking at a UC Forum. Month and a half away, he made a note to look into it, he thought better of it, and carefully took the flyer from the glass, or started to when he noticed a stack of them near the magazine rack. He walked over and took one from the stack.

He also picked up a Renter magazine when Megan came up from behind and asked if he was planning on moving.

"No, no, I was just curious what prices for apartments are these days. I may want to get a smaller place."

"Now that's backwards, most people I know want a bigger place. Why do you want to go smaller; you live in a studio now don't you?"

"I don't need any more space than to sleep, pee, and make tea in." This was one of his common retorts when asked about his living accommodations. Megan at least had a TV, computer, and microwave. To Hank, these were possessions that only required maintenance and distraction. He always wanted to find his way; his Spartan lifestyle was conducive to that. Hanks apartment was empty but for a meditation mat, a sleeping roll and minimal cooking utensils; a teapot, a pot, and a pan. There were no pictures on any wall. There was only one vase, a small 4-inch white graceful vase on a counter in the kitchen. It had one red Gerber Daisy most days.

Hank said that it was the most beautiful flower in the world; it was simple but displayed an elegance that he marveled at. It was representative of how he hoped himself to be, simple but complex in an elegant way. He was childlike in his view of many things, Megan had come to see. She knew someday there would have to be talk of relationships, love, sex, and all the trappings; but it had not happened yet. Hank seemed to be oblivious to these.

"I know there is a studio in my building for rent. Funny, but it is too small for most. The manager said he couldn't rent it because it is only 400 square feet. He was talking about making it into a storage room. I didn't know you were looking. I could ask about it," She said.

"I was not really looking, but since it is an option, I could take a look at it." Hank said.

Megan hoped the segue into living closer together would have just popped out of his head, but wasn't surprised when it didn't. "I'll ask him tonight when I get home." She said cheerfully.

"Did you know Swami Koan is speaking at a Forum in Berkeley next month?" Asked Hank.

"No, I would really like to go to that. You want to go together? He was so good last year!"

"He was enlightening, yes. Yea, let's stop in at the campus box office and see how to reserve some seats."

20

OK, I have to give a confirmed thumbs up on this equipment, there are no electronic failures, what you're seeing is what you're getting." Said Suzanne. She had spent the last week calibrating, testing, replacing components , and providing redundant recording devices so there was absolutely no doubt about the measurements. There were dropouts when Karen stepped onto the net.

"This is so strange, I don't feel anything..."

"How could you?! Those are cosmic rays, protons, they are relatively massless; you could never feel one or ten billion when they hit you. In fact, just because we aren't recording it, you do know we are being pelted by them as we speak."

"I know, it's just...they're stopping in me. That's scary, why are only some of the particles stopping?"

"I don't know, but we can be assured that the recording equipment is operating at 100%. I have also increased the resolution of the net by a factor when we added the new net. I'm now able to determine where on your body these particles are stopping. That should give us another clue we need to figure it all out." Suzanne continued. "Stop looking worried, this is probably normal. Maybe it's ET and we should have SETI call." She half joked.

Karen was worried that the experimental net was somehow doing something to her and her baby. She had recently gone to the doctor for her second trimester checkup and though she did not say what she had been doing, she asked the doctor to use care when he did the ultrasound. "No, there was nothing wrong with the baby, everything ball in circle, right where they should be," Said the old Navy doctor. It was his way of saying that like the fighters coming in for a landing on an aircraft carrier, when the ball is inside the circle, the pilot is on the correct glide slope to catch the arrestor cable.

But still Karen worried. She dutifully recorded her brain waves to record changes in the wave form so she could compare them to her recorded moods, stresses, illnesses. Her worries were elevating her levels of beta waves slightly and she discovered less REM sleep, but nothing outside of the extremes she had been recording for the past eight years.

Standing back and looking at the whole project, she saw this as an important breakthrough. Never in the journals had she ever read of anyone actually noticing, let alone recording a dropout of energy. She was pondering if she should publish now and keep studying lest someone beat her to it or if she should gamble and come to more conclusions to populate a theory with more data. In her mind she knew she would gamble, that once Suzanne had confirmed the recording devices and the electronics, then recorded studies could be repeated and she would collect enough data to postulate what it meant.

For a week, the theories were unspoken until Suzanne had confirmed the equipment was performing correctly. Now that she had demonstrated it was repeatable, redundant, and stable; even JJ began to speculate. Karen and Suzanne both began to speak, Karen asked Suzanne to go first.

"I think the particles are always there and just like in heavy water, the particle is absorbed when it hits a deuterium atom head on, something in your body is catching the particle."

"There are probably some atoms that are in the human body that will do just that." Said JJ. "You guys are the scientists in the family, what do you think?"

"I have been thinking about this since the first dropout. We've monitored thousands of frequencies and finally found a range that demonstrates some variation. We know when Karen is on the net, there is a shadowed effect, the particles yet at these high frequencies, these particles tend to pass through everything; through the roof, the bed, me, the kids, I even put the pet squirrel on the bed as a test."

"We know there are no other frequencies that dropout; only those coincident with the cosmic background radiation. We can measure the attenuation as a percentage of the net area before and after and we make the net a grid to identify with more accuracy where specifically in the grid the dropouts occur."

Karen was listening to Suzanne in rapt attention, as they had been doing all week, believing that repeating the simple facts could spark more thinking of what the meaning was; if any. But even if it was a natural occurrence of some kind, this was still very exciting stuff. This was new knowledge, nothing in the journals, nothing in the books; never known about. Karen was again off thinking how this could be compared to an asteroid hunter; or a comet hunter; finding something in the heavens that only you can see. She passed that thought and went back to her initial problem; how does the brain wave work, what is the initial spark, how does it keep regularly producing a modulated wave, how can I modulate it externally, how can it be stimulated into correction when it goes awry. That is what I started out to do, is this related or not? Am I looking at the right things?

JJ and Suzanne were both looking at her as though she had just come out of an unconscious state, "Karen, Karen—That must have been some thought. It was like you weren't here for a minute. What was on your mind?" JJ asked.

"Sorry, lots of thoughts, namely is this important or is it going to help with what I started to do with this project. I'm constantly worried I won't produce any result and the grant money would have been wasted, it could have gone to better projects, ones that could have helped people."

"Stop right there! Do not go on with that thought! We have no clue if this is useful or not. It is interesting and it may not be toward the end you are looking for, but that is no reason to believe any research is not helpful! Your vision is driving this and with what we have accomplished so far, you could write two books in addition to your notes and research reports. So, get back on track, put your scientist hat back on. Here are the facts, what can we postulate?"

Suzanne had to give this pep talk to Karen many times. It is discouraging to look for such a small thing as brain wave variations and hope to find a way to externally modify them to cure illness, but if no-one was looking for the answer, there would never be a treatment. Karen went through the ups-and-downs of research depression, all researchers do. Hoping to find the answer and every test yields no clue, it can be discouraging. Together they carried on as they had in the past and predictably Karen said, "You know, you're right. Let's figure out what this is and what it isn't."

21

A phone rang in the room of Dr. Lester Warwick. He was sleeping with his shallow breaths blowing over the partially closed journal he was reviewing. He had chosen to stay at a local motel, not a brand name hotel which he was accustomed to. He was trying to be incognito; he was not expecting any phone calls. The phone rang for the second time; he was now almost awake, he rubbed his eyes and stretched his arms wide and was remembering where he was again, why he was here and by the third ring was beginning to go through the realization that he had given his information to no one. He took a drink from the scratched bathroom glass cup, the water clearing out the residue of a short nap. It was refreshing; he cleared his throat and by the fifth ring decided it could only be the front desk with information that he didn't have to worry about, his whereabouts being known. At least that was his hope.

Lester had entered the camps at Bergen-Belsen; he had seen the pregnant women first hand. He had been successful in bringing medical aid to them; all he had done was positive. But he had not known that the US Government would also be so interested in the research started by the Nazi's that his life would change. He had known his every movement since returning from Europe was followed and his contacts recorded. It was something he did not expect after his research into the energy experiments he discovered after the war. It was all Lester could do to keep his postulations away from the prying eyes of the OSS and then the CIA. His fear was them concluding what Lester had now spent many years believing; the horrible ramifications if they decided to pursue that line of research and find the truth. A truth that could devastate the world in a fell stroke. A weapon which would end civilizations without the loss of a tree or the decay of a plutonium ion. Lester knew if the truth were known, the world had a limited life, and he had dedicated his life to following the researchers on a quest, with good intent, to prevent the discovery during his lifetime and pass the knowledge on. He knew it would lead the major powers to a climatic end if the truth was out. To his favor, there were hundreds of experiments they discovered and unraveled, many so far fetched they were dismissed as impossible. This might have been one of those, but still they followed him, after all these years.

He answered the phone with his groggy, gravelly voice with a simple, "Yes."

To his dismay, there was silence. After fifty years, he knew what that meant. At this second, the NSA Voice Recognition unit had analyzed the groggy voice and was concluding that Lester was in fact in El Cajon California, at the *Little Inn* on Mission Gorge. He knew that not answering the phone would have brought the more direct knock at the door from a surveillance verification team, so not answering was his way of buying more time. He knew from practice, he could actually lose himself in the next 10 minutes and have some privacy, but he had to move quickly. He packed his small overnight shaving kit, his books, and his glasses and made his way to the rental car and shortly found himself in Chula Vista where he hoped to disappear, because he could not risk them finding what he knew and how close some of his colleagues, like Karen, were to a breakthrough. Lester had several researchers he kept tabs on under the guise of grant reviews or reviewing the trade journal articles for; and on the rare occasion, visiting them personally to make the contact under the guise of professional reviewing and for old times.

Karen had submitted a grant review report of her research which Lester had read, knowing that on her present course she would find the answer, soon. He was in California to find a private moment with her, to explain what he knew and warn her of the danger. He dreamed of the first time he would have this conversation with a researcher, what he would espouse would be considered the rantings of a madman. He worried for years. It was his good fortune to have seen Karen was close and he knew her well enough to talk to her. It was thwarted in the parking lot by the company she was with. He could not bring himself to mention anything to Karen with Suzanne there.

Now he had to plan how to find a meeting with Karen that would be more private; a meeting he knew could change the course of history in the world. But in the meantime, he had to steer the government away from the work Karen was doing.

22

Hank and Megan wandered around Berkeley Campus for several hours, stopping for a coffee on Telegraph, laughing at the Tarot reader clients and gauging their fortunes by the expressions when the reader spoke to them. It was entertaining and they enjoyed people watching with the backdrop of Berkeley and the tremendous diversity the college provided. From the homeless to the executive businessmen, albeit long-haired, there were economic extremes rubbing shoulders. There were also the artists and the militants, young and old. Berkeley was a draw to all walks of life and being a resident provided many hours of enjoyment if you took the time to observe the passersby.

They wandered east on Durant Avenue to College Avenue, by Frat row, in the shadow of Cal Stadium. They spoke about the upcoming forum; Hank had bought two tickets to see Shaolin Koan. He noticed a post on a bulletin board that the abstracts and a website for more information on the forum papers and subjects were available. As he read the post, Hank felt a quiver of some intuitive energy when he read Karen Jordan's name. He recalled she was working on energy in the brain, studying possible modalities for dysfunctional brain wave therapies. This feeling surprised him.

He pointed it out to Megan and asked if she could get on her computer and read up on the forum presenters. Megan jumped at the chance to have Hank see her apartment and more importantly have Hank in her apartment. Maybe this was his way of flirting, maybe all he needed was a reason to find privacy, she was thinking in overdrive now. Stop thinking she told herself. He is just interested in using her computer. She said, "Sure, why not now." They strolled west to cross Telegraph, back to her apartment.

They passed People's Park, now filled with rowdies and weed heads, ecstasy sellers, and the hard core drug dealers. The center of the Berkeley drug trade, in today's world, this park has probably more web cameras in it than any other. They crossed Telegraph and talked about his curiosity with the feeling he had when he read Jordan's name. Megan read it and had no reaction at all.

"Have you ever heard of her?" Megan asked.

"I don't recall if I did."

"Maybe it has something to do with the energy sensations you've been having," Megan replied.

"I don't know; it's not like it's that clear. I don't think I've ever felt this so strong."

"What do these energy sensations you see look like now?" Asked Megan.

"I see almost an aurora, then a flow of energy, as though small amounts of energy are pulled into one focus, creating an intensity that shows up like a stream of flow."

"How do you suppose the speaker fits into this?"

"I don't know, maybe in no way at all."

They turned the corner and walked up the two flights to Megan's medium sized apartment. It was simply decorated, clean, and tidy. It had a pleasant smell of honeysuckle which wafted up from the inner courtyard on the U-shaped building through her open windows. The garden in the middle had a sculpted fountain with the sound of flowing water. It was a delight to the renters and though in a college town, the renters here were all-year-round tenants. This was home and they kept the garden immaculate and planted with beautiful flora.

The pleasant aroma brought Hank back to China for a moment, a familiarity he instantly felt. He looked around the room as Megan hung their jackets in the closet. He saw simplicity and complexity in the colors and choices of decorations she had. He noted nothing garish or offensive to his liking and this reinforced his comfort level. He saw her small desk in a corner with a laptop and mouse and asked, "Mind if I go online?"

"No, get started and I'll make some tea."

Hank sat down and found the on button, pressed it and watched the colors of Windows glow to life. He was quickly into Google and did a quick search of "Karen Jordan" just to see what was there, and if he had heard of her.

She had written several articles, no books, had a degree in Psychology and Electrical Engineering, did a residency in San Francisco, started a private practice and gone into research. Her specialties were Electroencephalogram use and she had written her last article on the use of external magnetic frequency control for stabilizing aberrant brain wave activity. It was hoped that using a non-drug therapy and electrical implants, similar to a pacemaker for the heart, severe dysfunction could be controlled giving patients a better future and brighter outlook without the use of drugs; not to mention the societal effects of not having violent dysfunctionals walking around.

He read further that she had spoken on several occasions, but never in China or Berkeley where he might have been at one of her presentations. But there were no references that came to him other than the intuitive feel that he knew her.

Megan came and kneeled near the chair to have a view of the computer screen and asked "What'd you find?"

"I looked up her name and saw the abstracts of some papers she had written, but I don't believe I know her. Funny thing I have this feeling, it's the same as earlier, and now this reinforces it."

"There is a purpose for all things," Megan said, reciting old proverbs. "I guess more will be revealed. Since you're over, how about dinner."

When he looked at her face, it was soft, feminine and again he had a strong feeling of familiarity. He wanted to touch the soft visage, push the hair aside, and touch the warmth that seemed to emanate from it. He was surprised by the feelings growing inside him.

With as much devotion to the Tao as he had spent in his

life, he neglected many of the social functions. Now, he felt out of water in this situation. In truth he was attracted to Megan from his first meeting. But he had never been romantic with a woman and he didn't know what he was missing or how to develop a relationship.

Megan sensed the attraction and was waiting for a sign, for him to say or act. She had been in this situation before and always felt a little deflated when he did not pursue any further. She was determined not to let the opportunity pass.

She looked into his eyes and said, "The way of the Tao is not required to be a lonely one. There are many couples in the ashram. I've known you for such a long time and have always thought we share many things."

Hank listened and wanted to hear Megan say to him that it was alright to touch her, to enter into a more intimate area. He was aware of the lack of relationships in his life, but it had never presented any pressure, never came out of the disciplined bottle; but he was beginning to feel it opening. He had an urge he could not control and he was listening so close to her every word that with all the thoughts in his head, he had entered almost a meditative state.

Megan touched his arm and brought him back to her room, her feel, her breath...

"Are you OK?" Megan asked.

"Uh, yeah, uh, just was lost in thought."

"Please tell me what you were thinking," She asked, hoping for, longing for; it to be her.

"Megan, I hope it doesn't offend you, I was thinking about touching your hair and face. I'm sorry, I should go..."

"Please do, touch my hair and me..." Was all she said.

Hank was relieved, he was excited. He was clumsy and brusque at first until his nerves calmed. He took her face in his hands and stroked her hair. He had experienced erections before, but this time it was very different. He was surprised by his body's quick reaction. She motioned for them to go to her small futon and they strode over with their arms interlocking each others bodies.

Hank gently sat and Megan straddled him, her legs folded on either side of him; they kissed for the first time. They were both so excited that the passion overwhelmed their sensibilities. Hank was outside himself, he was delirious with the touch of her, her mouth on his, her tongue massaging his. He exploded in feelings never felt and was overjoyed by the emotions stirring inside his mind. As he swirled in his ecstasy, he couldn't notice Megan in her overwhelming emotional state. She was flooded by waves of pleasure, longing relieved, she felt bursting inside as she allowed the flow of passion to totally engulf her. She longed for his touch for so long. Their body heat was multiplied by her breasts touching his.

Their first kisses were passionate; Megan was lost in the feeling. His lips were fiery warm, their tongues touched with warmth she couldn't describe. It was the epitome of longing end, of craving sated; of loneliness departing. Her body warmed and began to heat up from deep within her, heated by the passion of being touched and of touching. The comfort between them was so complete, they fit into each others arms with such ease, she adored the feelings and wanted them to remain forever, to live this ecstasy as long as possible.

Hank was indescribably content, though a novice, he was not unaware of love and sex. He was so moved by the intensity of their force, their desire, and her openness to his exploration of her body. When they were both naked, having made love, laying on the futon in each others arms, he felt the comfort of a lover having satisfied a long missed act of total intimacy with a woman he liked and respected. Megan cooed with a contented bliss, only wishing that it could be every night, but overjoyed for the success of the first time.

23

Hans checked his voice mail and was relieved to have the American reply so quickly to his request for a seat at the Brain Function Forum. Unusual to be so efficient, he thought. He checked the name again, Hadlyn, it does not seem German, though he was efficient.

He opened the trade magazines again and was perusing through EEG specifications and articles from practitioners for any hint of a discovery of some new kind of energy. The last time he read anything remotely close was by a woman named Jordon. He signed up for the conference because she was a speaker. Perhaps she had discovered something interesting.

Her abstract hinted at the utilization of monitoring devices to establish a baseline of wave patterns feeding back into a wave generator that would suppress abnormal wave patterns. The idea was interesting, but was it feasible? Hans wondered if in her research she had uncovered anything useful to his cause. He decided to write her to get an idea, if she would reveal the state of her research.

He wrote:

> Dear Dr. Jordon,
> Good day. I am a clinician in Bremen, Germany and was pleased to see your name on the upcoming forum in Berkeley. I write in hopes your research has revealed some effects which I am also researching. I have been in search of methods of utilizing energy waves which may influence certain brain activity. I have followed your work for a number of years and plan to be in

attendance at the forum in San Francisco. I have read your reports in the journals with much interest and it seems from your abstract there may be some new techniques in store for us. Could you send me information the specifics of your latest findings? It is my hope we could collaborate to some degree for the betterment of mankind.

Regards,
Dr. Hans Schick
hans.schick@bremen.isp.de

"If her abstract and her last article are any indication of her potential to discover the energy, I will be one step closer to my goal." He thought. He proofread the letter, penned it will his quill pen, allowed it to dry and placed it into the addressed envelope. It was addressed to the firm paying for the bulk of her grant. He was hoping to find her home address; he had plans he had to make.

He returned to his study of the pregnant Bergen-Belsen women to find the clue why they were special. He returned to some of the reference materials tracing the locations of his grossvater and other doctor's assignments. He read in some of the Nuremburg documentation that one of the crimes against humanity was the experimentation on live humans. Ha, he thought, they were not human...In all the Bergen-Belsen paperwork, there seemed be to something missing. He knew of rumors of an American doctor who took the documents before the officials confiscated them. He always wondered if that were true and in the course of many years, he was able to prove the American Doctor was Lester Warwick. He and some Aryan funded colleagues had been following the good doctor for some time and also knew the American OSS had put him under watch. Then after they became the CIA, he was still on the follow list.

Hans had sent another request for information out to his investigative friends and was awaiting the coded report.

It had been so many years, regular weekly reporting had long since turned into annual reporting unless a special request was submitted and approved. Hans was headed to the United States and wanted the annual report of Warwick's activities early. He could investigate further when he arrived, since he had a reason to be there to cover his true intentions. He had long since given up the idea of getting any documentation Warwick had taken; he had never revealed any papers that were useful. Since the US Government followed him, an occasional Freedom of Information Act request for newly released Nuremberg information always came back negative. But there was always the chance that something might make Warwick pull the papers out of his hidden closet; maybe allowing Hans to get them.

He emailed the coded request and returned to the boxes he had recently checked out of the Bundes-Bibliothek, launched his Wagner DVD, opened his wine, and prepared for the next three hours of research.

24

Getting the second net ready was taking much longer than Suzanne anticipated. On paper, matrixing the grid looked like a piece of cake, but the small energies they were measuring were hampering her efforts at calibration, once she could get it to record. All told, she spent two weeks to get it to the point they could even establish the calibration level of energy that contained the dropouts. She was disappointed in her results and the time invested.

Karen said, "I wish you wouldn't be so hard on yourself, it is so unlike you."

"But I hoped to have this done long before now; I feel I'm holding up the parade. You have the conferences; we are dying to know what the dropouts are..."

"Look, you are awesome, your enthusiasm has always been equal to the success we have had already, and there is enough now to get me through the grant review process. I really think it shows that regardless of the cause of the dropouts, external energies are definitely affecting the human body; and if something external is at work, we can use something similar. Don't worry about the forum and the presentation, just because we think we are close is no reason to not take all the precautions to do it right. Be the perfectionist you are and don't be so impatient."

Suzanne thought better of her own judgment before this, but she ceded it wasn't the end of the world. "I know you're right. I think we can get the system back online today. I hoped for last week. One thing is; it'll be right. I've learned quite a bit on the neural netting. There is some fundamental sinusoidal canceling going on I spotted earlier. I spent a long time the first time working through them, this time it was only two days of compensation. My little success story, more a technician's pride thing than relevant enough to report to you about."

"That's awesome, but you should've told me, I'd have added that into the paper. I'll get on that for the next paper. So it looks like we'll be up, today?" Karen said.

Just then JJ brought the boys up to the office and asked if there was a chance of getting them out of there for a walk. "I know you are really in the middle of things, but it is a beautiful day. How about a hike?"

"It looks like we are going to go live today. I love the offer, but I'm getting itchy to see how the new system is going to work. Come here you two!" Karen said and the boys raced over to her.

Stephan showed Karen a stink bug and said, "it tickleth my hand when it walkth!"

Jeremy said it wasn't that big a deal which drove Stephan nuts. "Stink bugs don't tickle, tarantulas tickle!" He said, knowing that Stephan was still leery of the large hairy spiders indigenous to Julian.

"It ith too! You don't have one; tho you're jealeth!"

"I can get all I want..."

"OK you two, let's get on the trail, and try to find a garter or a king snake," Said JJ, trying to stop the brewing argument. "We'll give you a couple of hours of peace Hon, be back around 7," And he kissed Karen before leaving.

Karen wanted to go on the hike, but she really wanted to see the new neural net operate. "See you then, I'll have dinner made, spaghetti sound OK?"

Both boys agreed quickly and hugged Mom and Suzanne as they left.

"What are you going to do with the German doctor's request?" Asked Suzanne.

"I was thinking of emailing him back and ask him to be more specific, I'm not sure what he really wants. I don't really want to get anyone's hopes up before we have something more definitive, I think if I talk about some of the results before we are ready, it would send the wrong message."

Suzanne asked, "Why would someone you don't know want to share our results? What has he written?"

Karen flipped her computer desktop to Google and did a search for "Hans Schick" + "psychiatry" and found nothing. "No papers, no research results, on the International Doctors Registry there are fifteen doctors with the name Hans Schick, one in Bremen. The city is all that is available, can't even be sure if Schick was the same doctor. Even if he was the same doctor, there have been no papers by him. I wonder what kind of research he's doing?"

"There's no history on him, he's probably just a hopeful doctor, hoping we can come up with a method. I guess I'll write him back and ask what his research entails, I'm sure it is no where near what we are doing." Karen said. "Maybe he is a rich cute doctor who would fall for a research assistant!"

"Probably an old fart with an end of career desire to publish something to be remembered by," Said Suzanne. She had a fling with a professor once and swore never to date older men ever again.

25

Lester was in Baltimore, visiting another researcher at the Baltimore Clinic. It was his way of keeping people from questioning his intentions and constant travel. Since leaving Santee and his meeting with Karen and Suzanne, he had been to his Novato California home, then to Denver and was wrapping up in Baltimore when he made plans to return to Julian and get the moment alone with Karen. He was at the Baltimore clinic as a visiting expert to oversee the work of Dr. William Cronin. Bill was 54-years-old, third generation American, and not doing anything remotely near brain energy research, but Lester reviewed them all. His EEG work was of interest as he was a test office for one of the manufacturers so Lester was always keeping an eye out for the newer technologies.

Bill and Lester went back many years, another colleague from Lester's early years. From the Eastern Shore, around Cambridge and Easton, Bill was the son of a tobacco farmer. He had gone to Harvard without an idea of what to do and found his calling in psychology, later in brain research. The Baltimore Clinic hired him to begin their Brain Study Clinic and after years of staff duties, became the Medical Director for the department. He and Lester had often collaborated on coast to coast conference calls, comparing maladies and prognosis of their respective patients. In fact, Lester and Bill had collaborated on a book about brain injuries. It was a regular visit to see Dr. Cronin in Lester's travels. As he sat in Bill's office waiting for Bill to return from meetings, he took the time to make a record of events of the week, especially the latest visit to Santee.

Lester refused to write anything down unless it was

encrypted using PGP, an encryption program commercially available on the internet. He had set several explanatory emails and attachments to be sent in the event he did not log on. They would auto deliver if something were to happen to him and a week elapsed without logging into his email program. It was his idea of not letting the information he possessed get lost if he were to become incapacitated. He was searching for the right person to pass on his knowledge to; Karen was the most likely, Bill Cronin another. Lester felt that the work could not be lost, it was too important for billions of people who could be at risk if the technology were in the wrong hands.

In the master plan set forth from Adolf Hitler, down through the SS ranks, any weapon to rid the world of non-Aryan people was fair game; and one particular experiment included identifying a way to stunt the growth of Jewish embryos. Lester's research included background data on a Doctor Schick, a Nazi doctor who had served his tour of duty during the war in Bergen-Belsen. He discovered in the camps, the experimenters could deliberately induce babies to be born with little or no brain function, in essence brain dead; without the use of any drugs or physical contact. Though they did not understand how the effect occurred, they found that energy fields could deflect away or produce some internal effect in pregnant woman which assured the babies were born unviable. Left to develop the technology, the Germans could have wreaked tremendous devastation to the human race. Fortunately the answer was incomplete and the war ended when it did. Lester was the only person who had the experiment documentation. Though illegal, Lester felt the limited success of the experiments would be better off not being made public or in any government hands, he removed it from the camp without reporting it to his commander. Then later, he denied it. He had flirted with treason, but his convictions were firm and resolute.

He shuddered to think of the weapon that could stop human development, in the wrong hands; no one could be trusted to use it wisely. The catch was, not even Lester knew what the process was; only the documents he stole away so many years ago had any suggestion. Something to do with brain energy; he dedicated his life to finding it before anyone else did in order to stop the diabolical concept from being explored further.

When he discovered the documents, he was disgusted by the brutal acts on paper. There were many papers that were nightmarish schemes of human destruction, so far out that anyone reading them would have guessed they could not be true or possible. It did indicate the depths the Nazis were willing to go to, to eliminate everyone but their own race. There were hundreds of ways outlined to eliminate human beings on mass scales. It infuriated him to read how casually these monsters wrote of life and death. Their lack of ethical concerns was emblazoned by doodled thoughts next to imaginary rays that disintegrated human beings that someone had drawn in the margin of one document. But the ideas that mattered to Lester the most were the documents of the experiments they were in process of performing involving brain energy and the results; brain dead children.

Lester had confided in no one about the papers and only a very few people about his personal philosophical concerns within the small group of EEG professionals about technology and misuse of it. He kept a vigil, looking for anyone who might have known about the wartime experiments and he kept keenly aware of any reports of brain dead births. He followed up every one and became known for his prompt care in the scientific study of premature brain death. But so far, none were the result of obvious efforts using a strange new technology. Lester would not rest though, in search of the weapon that could end entire populations without visible or physical intervention.

His interest had piqued when Karen began her research to design a modality using an external neural network for brain wave control. Because he personally knew her from her residency, she was not on his list of suspected neo-Nazi murderers; but he checked her progress. He maintained a relationship with her, though distant, he followed her research like a hawk. He was also on the advisory board for the *NEPA*, the *National Electroencephalograph Practitioners Association*. He was a strong supporter of Karen receiving the grant, though she would never know how much he influenced the decision. Lester felt if Karen could find the energy before anyone else, he could rely on her not having any evil intent.

His military activities at Bergen-Belsen had alarmed the US Government. From his service in Germany, he was suspected of taking important documents which he denied. He was watched around the clock. His initial error was in talking to his Commanding officer about his concern based on the documentation they were pulling out of the camp. It could be some of these techniques were loose in the populace and the government should make an effort to track the families of the doctors working on these projects. He was ignored. He wrote to his congressman and was muted, and the third time he tried to detail his ideas and concerns to OSS; it was then that he was ordered to cease to write about these things. He wasn't being heard and was told he was spouting inanities; chicken little, brain-death from an antennae, it just wasn't possible.

When he could not convince them, he copied one non-specific page of the SS report of the findings in Bergen-Belsen as proof of his assertions and it was that document that got him in big trouble. If he had not taken any documents, where did this one come from the government asked? Lester was torn between his belief that turning them over would expose a great weapon to men who might use it, and the personal risk of prison for stealing official documents and then lying about the fact. He opted to keep the documents secret, but his life was a constant series of personal invasions. He would come home to overturned furniture; looking for something but not even the government knew what it was. Certainly there were documents, but how many pages? What other details were in it?

Lester had long since encoded them and safely stored them in the ground, in a bottle with a waxed lid under the gravestone of a coded grave near his father's, 1945. One right, nine left, four right, and five left. The year he found the documents. The year he began his relentless search for the technology so he could stop its deployment.

He needed to talk to Karen without the government listening. He knew over the years he must have slipped up, by mentioning things that in context could only mean he did have some knowledge that could only have come from the missing documents; but he was getting old. He needed to know if Karen could be trusted and if she was close to the breakthrough. Her abstract for the upcoming forum had led him to conclude there had been a breakthrough of some sort. He had to know what she discovered and tell her the truth of his concerns and to warn her of the danger she was in.

26

Karen adjusted the interferometric field and was able to pinpoint the area of the net where the dropouts occurred to within .05 mm. Suzanne had been clever in the matrixing hardware; with test objects, they could identify very precisely electromagnetic disturbances. The interesting thing was now a wait and see, when the dropouts occurred.

Karen laid on her bed to take some test recordings of her neural activity with Suzanne glued to the output displays.

"All normal, the net is detecting comparable data to two weeks ago; at least we are back to where we were before the additional matrix was added." Said Suzanne.

"Is it possible to amplify the individual waves, isolate the alpha from the beta?" Karen asked.

"I can amplify the entire spectrum, I can isolate by physical location, hold on... Wow, I didn't think of this before, but with this kind of discrimination, I can pick the two-dimensional plane of the brain activity and get very specific wave patterns!"

"Unexpected plus, I never thought about the ramifications beyond the ability to locate the dropouts. That may be very useful for diagnosing aberrant wave generation. It would certainly help in diagnosing traumatic injury contraindications." Karen said as she was now considering the fortuitous outcome because of a setback. Research was disappointing and then spectacular.

Karen continued, "So if we can get a two-dimensional spectrum, we could add layers and generate a three-dimensional image of misfiring synapses, record where they are and record the wave form and analyze the causes. This is what we have been looking for!"

Suzanne ran over to Karen for a hug and a brainstorm session on the next step when the wave monitor went into alarm.

"What's that?" Asked Karen.

"Not a problem, I added an alarm to allow us to know when the dropouts occur so we wouldn't have to wait for the nightly run. Wow, it is definitely happening right now."

"Where are the dropouts happening?"

"It is abdomen...Lie still...Damn, I need to overlay some reference to your body. Spread eagle, point your arms and legs to the bedposts and center yourself."

"OK, done. What does it show?"

"It is plotting on the net at X=44,528,998,552 and Y=53,444,599. Don't move from now on, I will get a fix and transfer it to your body, but I will need a minute to get it done. We are recording, this is fascinating. There is a zone, a small area, not just a pinpoint where the dropouts are happening... Stay still while I get a tape measure."

Suzanne ran out of the office to the kitchen junk drawer for the only tape measure in the house. Mostly used for picture and curtain mountings, it was a Home Depot orange 12 footer that was as good as new.

Suzanne ran back upstairs and was interested to note the dropout alarm was still ringing and she began her measurements over Karen. With some careful plotting, Suzanne looked Karen in the eye and said, "It's over your womb."

27

Hank and Megan had now spent the better part of two weeks living together. They were thrilled to have changed their circumstances, to have a soul and room mate, sharing everything. It was as though they were alter egos for each other. Megan wondered aloud if this were too much of a coincidence, they both were practicing Buddhists, both worked at the same job, interested in very similar things. Was it too much Megan worried?

Hank was thrilled with no trepidations. He had not thought much of loneliness, but having a relationship had juxtaposed the times alone in a very powerful way. It filled his mind with the urgency of maleness, the desire to make love whenever the urge struck, to have the option to ask and not worry about rejection. He had found something missing in his life; intimacy with another person.

He wondered if he had selected the way of the Tao to explore being alone to avoid noticing he was alone. Megan and he spent many a night discussing what their purposes were, what brought them to meditation, and how quiescence is hard to obtain, harder now that there were these new forces in their lives. Did they know what they were exploring? Were they prepared? How could anyone be prepared for such longing to come to an end?

They laughed and cried as they poured out intimate stories, neither one holding back, finding synergisms relating more personal stories. It was such a joy. The newness had yet to wear off and neither of them wanted to leave. They worked and meditated together. Hank was having difficulty getting

meditatively lower because of the constant thoughts of Megan. Her scent across the room caused his mind to wander. He wondered if this would change. He knew it must, but he enjoyed the thoughts of it taking a long time.

They decided to meditate at different times; Megan was having the same difficulty, unable to clear her mind when Hank was in the same room. She was content, such a simple word she thought. I don't think I've ever felt so content, she thought. For the first time in her adult life, there was someone she respected and liked and it was now bordering on love. Their day at the ashram was spent separate but still together and they were planning the evening.

"Hank, let's just order a pizza and stay in tonight. It's drizzly and I just want snuggle with you."

"Then it's at your apartment. You have the better heater and I like it there all the sudden!"

"Want to get a movie?"

"No, I just want to curl up next to you babe."

As they rode home, the streets glistened in the light from the streetlights and the occasional headlights of passing cars. Down University Avenue, not stopping for their usual tea at the Marrakech Restaurant displaced now with their own teapot.

Their diets and their haunts became jumbled into what was once a feeling of dinner with a friend to dinner with a lover. They were noticing it was OK to withdraw out of their usual outside stops and find each other's company in the privacy of their apartments. They were giddy and both knew what the other thought. When one laughed, the other knew why. The Bhagavad-Gita spoke of love and its mysterious power, but this was amazing to them. They got to Megan's apartment, changed into dry clothes, kissed, and hugged as they plopped down onto the futon.

"That forum is coming up next week; I was thinking we could make a reservation at the Durant Grill afterwards." Said Hank

"Oh so special! What's the occasion?"

"I want it to be a special night; I want every night special.

Besides, it's where my Dad is going to meet us. He's coming in for work, but not in time for the forum. So I planned to meet him there after the forum. I want him to meet you."

"Why didn't you tell me?" Megan asked.

"I just found out. I was considering not telling you for the surprise, but I couldn't do it. There will be other surprises."

"Like what? What are you saying?" She tickled him.

"I'm not saying anything more." He took her hands into his, guided her arms behind her, and kissed her.

28

Herr Schick made his travel plans, arriving in San Francisco four days before the forum. He made plans to meet with friends of the Order. He heard from the Jordan woman but she only questioned his request. Even in this day of email and Google, he was not privy to Karen's information and it frustrated him that she had not responded with substance. If she would not offer her information freely, he would force the issue. He opened the report from his fellow operatives, the special request he made two weeks earlier. In it he found what he was looking for. Karen Jordan's address, vital statistics, family life, pictures, and briefs on her husband and her research assistant.

He mulled over a plan for over an hour; to take the next step was dangerous. He knew he had to be correct; to risk exposing the network of "friends" around the world would not help the cause. They had been successful in remaining anonymous and authority free, though they still practiced their skill at killing non-Aryans and promoting the correct people into power. He considered all he had learned, his experience with his neural net and the apparent key found by this Jordan woman. He thought it ironic that she found something before him; she is probably from German stock he thought as he raised his wine glass.

He considered the risks of losing to the scientific world the technology that seems promising. It would be much better to have it before anyone else can discover it, ideally if the woman is willing to sell it; that makes the matter an easy decision. But if she is unwilling, then longer strides should be taken. It is too valuable to lose. Yes, it is worth the risk.

Once again he read her abstract. It was general, but it was so on point to his conclusions of the necessary tools to complete the experiment and find the greatest weapon ever created. He was driven by his staunch training and unwavering dedication to fulfill the promise to his grossvater to find it. He was also driven by the dark need to return the Aryan race back to the top of the human species, by elimination of all others, if necessary. His grossvater had left many things for Hans, the legacy of finding the energy but also the memberships into exclusive organizations would one day return to dominate Germany for the glory of the Race.

Hans made a call to confirm his intentions to the *Order* and outline the advantages and disadvantages. His mental notes on the importance of this technology were communicated again; he picked up the forum pamphlet rereading the abstract:

Thursday 9:00 a.m. to 11:45 a.m.
Dr. Karen Jordan, BSc, PhD, MPsy, EE

Neural Energy Manipulation
For eight years, Dr. Jordan has been researching the use of external field generating modalities for prophylactic elimination of aberrant brain dysfunction. In this session, Dr. Jordan will explain the theory of using modulated reverse neural electro-injection to control alpha, beta, delta, and theta waves with a modality during sporadic outbreaks of waveform disruption during neurotic events. Her presentation will focus on the design and implementation of neural matrixing for extremely fine recording and possible non-invasive waveform injection techniques for reducing and possible eliminating the effects of severe brain dysfunction.
Dr. Jordan received her Bachelors of Science Degree in Psychology from the University of California, Berkeley, School of Psychology....

"Reverse Neural Electro-injection. If this is purely passive, then it is not what we are looking for. However, with electromagnetism, it is rarely one-way. What can be detected can be generated. If she has found the correct spectrum, the value of this technology is quite high." Hans spoke into the phone to *friends* in California.

"This technology could be very useful for our aims, we must not be found out. I will offer to fund her further experimentation provided she comes here, or we will take whatever steps are necessary to get this technology and leave no-one behind who can identify us. I will be in San Francisco on Sunday. We will discuss the details further when I arrive in country."

The voice on the line said "Ja."

He hung up the phone and took another drink from his wine glass. "Today the answer may be near," He toasted his grossvater's picture. He proceeded to reread the files, the pregnant woman from Bergen-Belsen. Why this woman? He concluded that his grossvater had only been scribbling in the gutter of the document, or it was the top file and the pile was assigned to him. Besides the obvious genius of stopping the race before conceived, what would they have been doing? In 1945, they certainly focused field capability, but even though such power was possible, the destruction of life would require a proximity that made it unusable on a meaningful scale. What was the true nature of the energy, how did it destroy life?

29

Karen stared out into the ether, eyes focused on nothing, worried. After a moment, her eyes refocused and she called out to Suzanne who left the room to assure the recording was continuing.

"Suzanne! What is that noise? What's going on?" Karen asked.

"I added the alarm when the dropout condition occurs and I admit it's too loud, but it is all OK. Calm down. Karen, there's nothing to worry about."

"OK" She said. "Where do we go from here? Talk to me..."

"I think we need to just wait for the dropouts to stop. History is full of four minute dropouts; we're only into the third minute. When the alarm stops, let's keep you still for a baseline again and then we'll analyze what happened."

"Why is it the womb? Why not the brain?"

"Another good question. Why are these frequency particles stopping in the area of your womb?

"What's in a womb...embryonic fluid, umbilical cord, the fetus...? When did the dropouts start? Seven weeks ago. I am 16 weeks pregnant, which means the dropouts started when I was nine weeks pregnant. Would the machine have worked before the first dropouts were recorded?"

"I need to check the notes, but we had changed the scanned freqs and only had nine days of recording before the first dropout. Either in the first nine days there were no dropouts or the net was unable to record them. We should be able to determine that. But what if it is related to pregnancy? How? Why?"

"I don't know. What is CBR anyway? Why is there so much of it? I need to know more about it before I go on."

"Karen, this could be the breakthrough, you can't stop here. We should continue recording, isolate the specific frequencies, and then isolate the energies for a start. Maybe we can do it from the recordings we have, but I know we'll need to fine tune the interferometric matrix."

The alarm abruptly stopped. Suzanne ran to the office and verified the recordings had stabilized. The recordings stabilized back to the readily familiar waveforms of a normal healthy brain and the previous level of energy. The normal amount of CBR entering its unique patterns onto the net, recording ten thousand samples per second.

"OK, I have baseline. You're a free woman." Almost instantly the patterns disappeared as Karen hopped out of bed and was at her desk, copying the data to begin a long overdue analysis of these mysterious dropouts.

OK, printing...Suzanne, pull the records for the last nine weeks. Let's prepare for an all nighter."

"OK, Boss."

At around 8 p.m., JJ and the kids piled into the house. Karen and Suzanne took a break and began dinner, while listening to the story telling and retelling of Jer and Stephan of a day's hike into the Cuyamaca forests.

"You look excited! What's going on?"

"JJ, something's significantly changed. Suzanne and I had our first look at some incredible data. We have no hypothesis yet, it is still too new to conclude anything, but the energies we are looking at have an absolute correlation to the womb; to my womb."

"Your womb?"

Suzanne piped in "Yea, the dropouts we've been seeing, they are over the womb area. It's like there are streaming particles all the time going through space, then for some reason, the stream focuses and the energy is stopping in Karen's abdomen, in her womb."

"You're kidding right? What does that mean? Is it normal?

Well I guess it has to be, unless the equipment is causing the effect...but the effect is unrelated to the equipment; right?"

"That was my concern too, that it's a side effect. But after looking at the results, there are no coincident test equipment events that occur when the dropouts occur. It could be something we are not measuring or are able to measure, but it is not the interferometer or the nets or the computers. There are no detectable coincident events." Said Karen.

The phone rang, Suzanne answered it. She smiled and said, "Oh Hello! Yes, here she is..." She cupped the phone and said "It's Lester Warwick."

30

God is the almighty, God is all powerful. God is the source of life and death. Only God can give life; if it is God's will, life will be taken. If you do not believe this fundamental truism, you are doomed to a faithless existence and heaven less afterlife. If you do not accept the Savior into your heart, your path to righteousness is blocked by the sin of non-belief and a sure entry into the depths of hell." Said the Priest on the radio. Hank was surprised when he turned the radio on.

"Blessed are the tenets of the catechism and those who follow in the path of God's light, for you are one of the 12 tribes chosen to return to heaven on earth."

"Why do you listen to this?" Asked Hank as he turned the volume down.

Megan said "I'm Catholic. Though it may seem somewhat of a conflict with my meditative practices, I still have hope that finding the Tao will include confirming the roots of my spirituality."

"Do you go to Church?"

"Not now as much as I did. But I still try to experience the faith, even if it is only the radio."

"That's something I didn't know about you. We'll probably have lots of these conversations."

Megan was worried, but it looked like Hank would be accepting after all. Not critical like most men she had known. "Hank, what do you believe life is?"

"Ahh, *The* question. Is life an animal or a vegetable; is it a force developed from the slime pits or gifted from heaven?

Who can really know Megan? I have listened to many leaders, spiritual guides who mostly know that they don't know what life is. They can guide you through it, they can relate experiences, and they can help you invest well; but I have yet to see the definitive answer to what is life."

"You were in China, what do they believe?"

"There are the western influences; many missionaries visited Japan, Korea, and China. They spread their word and sometimes it stuck. My training was not purely anything because that doesn't exist anymore. It is all a melting pot of beliefs based on forefathers who write about the path of life and how best to walk the path."

"But do you believe we are from slime or from God?"

"Serious talk before dinner. I don't know. What about you, what do you think?"

"Left to the church influences, I have been ingrained with the Father, Son and Holy Ghost, Immaculate Conception and Adam and Eve. So many people believe it, I feel comfortable believing it, but I don't honestly know. I think the comfort of the church for me is that as long as I believe what they teach, I'm in good company. No one challenges the beliefs in my circles."

"Then why worry about me hearing your radio tuned to a Catholic station? I sense some curiosity."

"Is this a challenge? Ok, do you honestly believe Pangu came to life pushing the sun away from the earth and became Earth?" Asked Megan.

"I have no idea. But there are no sects in China repeating it on the radio for the benefit of the followers. They know its myth and believe it to be metaphorical; at least that's how I always took it. Maybe I'll believe it when another Pangu is born!" Hank laughed.

"Hank, I don't know what to believe, but I do know I want you to be around in my life."

31

Lester rented a car at Lindbergh Field, loaded his bags, and drove to Fashion Valley Mall in Mission Valley. There he went into J. C. Penny's, bought shaving supplies, pants, shirts and a small canvas bag to put them into. He went to the bathroom, changed into the new clothes, dumped the old ones into the waste bin, and disappeared into the crowd. He found his way to the San Diego trolley and boarded the eastbound trolley, bound for Santee.

When he got to the end of the line, he went to a pay phone, called Karen Jordan, and began a conversation that would change both of their lives forever.

"Lester, so good to hear from you. Are you in town?"

"Yes Karen, I am. I need a favor; can I trouble you to pick me up? I'm at the Starbucks in Lakeside, the corner of Weld and Magnolia. My car has broken down and the rental car company says they cannot come for another three hours."

"Well sure. I'll be there as soon as I can. You said Weld and what...un huh...Magnolia. Got it. It'll take me about an hour, but I'll be there. Is everything OK? You sound nervous."

"No, no; I'm alright, sorry for the trouble. Karen, Come alone."

32

The next stage in development of the nervous system is called the stage of histogenesis (formation of tissues). During this stage of development, nerve cells (neurons) proliferate (multiply), then migrate from the center of the neural tube to a peripheral location, and finally mature by the growth of processes, myelination, and the development of contact with other neurons (synapses). Neuronal proliferation occurs between two and four months of gestation. Neuronal migration spans a period between four and nine months of gestation. Maturation continues for some years beyond birth." Read Suzanne from her Google search.

Suzanne began the search after getting the kids into pajamas and bed. JJ tickled them, wrestled with them, calmed them down, and finally all was quiet. With the hiking they had done, both boys went right to sleep. JJ always thought, "Better than Benedryl" When the boys played hard and went right to sleep. He went downstairs and found Suzanne in front of the computer in the kitchen nook.

"Whatcha looking at?"

"I'm reading about brain development. Karen and I are recording energies in the brainwave areas and since the womb seems to be absorbing some kind of energy, I'm reading about the developing embryo. I don't know what to look for, just reaching I guess. Remembering my anatomy class, what this means is that brain synapses form between eight and 16 weeks of gestation." She said. She looked at JJ and said, "I had a thought and looked up some information to confirm a conclusion; the baby is developing synapses right now."

JJ asked, "What does that have to do with the dropouts?"

"Well Karen said something that made me think for a minute. Obviously there is a viable fetus inside Karen's womb and the CBR particles seem absorbed by something inside the womb. Our choice of frequencies to monitor to find neural resonance pathway to help amplify injected corrections has been the longest part of our research. We have scanned for over five years, once we got our apparatus going and when we finally find one, there is apparently evidence of energy accumulation. The energy is measured on the net, then the level goes down, it has to be somewhere if the energy is steady state. My theory is if there is energy inside something, there must be an evidence of the energy accumulation. What I think I've just postulated is some kind of neural enervation in your child is from CBR; but not all, only specific particles."

"Wait a minute, you mean you are theorizing the neural energy, that is bringing the brain to life, comes from outside the body? From radiation? I have never even heard of a theory for this. Is it possible?" JJ asked.

"Well at this point, it is not ruled out. Her fetal development is coincident with the development of your daughter's brain cells."

"Daughter! Karen didn't tell me she knew."

"Oops, JJ, please don't let on that we knew. She wanted it to be a surprise. Sorry, I forgot myself in all this."

"A daughter, that's great! I know Karen wanted a girl."

"JJ, this is truly Nobel Prize stuff. If the evidence proves out, Karen has just discovered something no one has ever postulated. I'm going to try to come up with other explanations, but there are some obvious implications here that have some far reaching ramifications. I'll call her. She should know this. I bet she's almost down the hill. This should really surprise her! I wonder if she should consider modifying her presentation, the forum is next week."

33

Juergen Hoffman continued his surveillance of the Jordan house from inside his van. On the side was printed "Red's Carpet Service". He was a member of the Order, a Hayden Lake graduate, he had sworn to reestablish the Aryan race to its rightful place; supreme.

The members of the *Order* were nefarious and sometime outspoken, but the specialists were indiscrete and secretive. As their membership grew, the necessity for information was prioritized and members were trained in various covert arts. Juergen was one of the bright operatives and specialized in surveillance and communication. He received a call earlier in the day and began preliminary observation as soon as he could. He had not yet bugged the phones or entered the house; that was next; after he established a pattern of family behavior.

His information on the target was generic; address, family details, husband, two young boys, and an au pair. Mountain home, highly educated; with electronics equipment that cannot be damaged.

He had been on station three hours when he saw the Jordan woman leave in her car. She was gone about 45 minutes according to the log. Since the recording devices were not installed, he didn't know where she was going. It was at 8:28 p.m.; maybe she needed something from the store. He made mental notes to pass the time.

From his vantage, he could not see the downstairs at all, only one window in the second story, facing the front of the house. He had seen shadows pass the curtain, but the lights went out at 8:53 p.m. probably a kid's room he thought. He spent his hours Googling the Jordan's. She was a psychiatrist, had written papers on brain stuff. He couldn't fathom why he was asked to watch them. His training was never far from his mind, he was to do as he was told and the bigger picture might never materialize. He had been in the *Order* for 12 years and there many missions he never understood the purpose of, only believing what he was told; that it was always for the good of the *Order* and to return to the Aryan race to supremacy.

He discovered Jeremy Jordan was in the banking business, traveled frequently given the number of references to his presentations out of the country. That means the man of the house is likely not a real man—number one and he is gone a lot—discovery number two. This should make it easy to get in and prepare the house. Though he was trained in killing, he preferred his surveillance and the art of bugging phones, Ethernet, and the thrill of breaking and entering. He settled in for the night and awaited his replacement in the morning.

34

As she drove into the parking lot, she did not see him immediately though he saw her. Karen found Lester in a pensive mood; he was worried about something. She found him inside, at the table nearest the door with his back to the wall of sundries for sale. There was a newspaper on the table and it was evident he had been reading it from the quick folded mass in front of him. She wondered why the instruction was to come alone.

She went to his table and Lester stood and told her they needed to leave. Without explanation, he took his bag and walked outside the store to her car, looked around, and got in. This was too much for Karen to grapple with; he didn't greet her in any manner at all, and now she was alarmed at his behavior. She was diagnosing him as she got into her car.

"Lester, what is the matter?"

"Drive up 67, as though you are going home. I need to assure that we are not followed. Do you believe you were followed, or have you noticed anything strange in the last couple of days? Strange inquires, different people around your house?"

"No. Lester, what is going on? Are you in trouble?"

"Karen, There is something I must share with you, but I must be sure that we are not being overheard. Please drive until I signal you to stop and say nothing else."

Karen was now considering Lester had gone senile or something worse. She drove back up the hill. Lester looked over his shoulder, out the rear windshield; constantly on the lookout for headlights that might be a tail. As she entered Ramona, he motioned for her to stop at a small restaurant and indicated for her to park in the back parking lot. There they waited 10 minutes for assurance they were not being pursued. Lester opened his door and motioned for Karen to join him outside. They walked five blocks away from the car and found another restaurant, and went inside. Karen was concerned, there was nothing in her life that required this much secrecy.

"Have you ever been in this restaurant before?" Lester asked.

"No, what is this all about Lester; I demand an explanation."

"Let's settle in and I will explain. You must have determined this is bizarre behavior, I assure you I am all here, this is not some over reaction, please bear with me for another moment."

They went inside and found the table as far from anyone available who might overhear the conversation and Lester sat down. Karen was right behind him stunned at the thought of this kind of demeanor from Lester. He had never acted this way before, though the years had gone by. She was unsure if this was his latest norm or not, still sizing up his state of mind.

Lester began, "You have known me for many years. What you don't know about me is what I need to relate to you. It is in relation to your research work and the importance of you understanding how close you are to providing one of the most potentially destructive weapons that can be unleashed on mankind ever devised."

Karen sat back in shock. No, No she thought, what I'm doing is only for the benefit of man. "What are you talking about? You don't have any idea what I have been doing. I haven't published the latest results, how could you possibly assume what I'm doing is related to a weapon?"

"You're right, I don't know the specifics, but I do know you are about to discover something I've been watching for the last fifty years, hoping it would not be discovered, praying that if found there is a way to neutralize it. Let me start at the beginning."

35

JJ was beginning to worry. Karen had left some three hours ago and there was no word from her. He called her cell and got voice mail at first and then finally Karen called back. She must have gotten the worry in his voice form the message; she called to say something had come up with a friend and that she would be another couple of hours. Don't worry, I'll be home soon.

JJ knew something was up. Why didn't she say Lester's name?

JJ texted a message to Karen's cell phone "Are you OK, is your friend OK, tell me what I need to do."

Karen showed Lester the message. He said "Text messages can be traced. Tell him you'll be home at 2 a.m."

Karen texted the reply and Lester continued his story. "After I exhausted all the government officials and was called a nut case, I decided to bury the document in a safe place. Since then, I have been determined to follow the technology until there was a breakthrough. Then, I could extinguish it. It was good fortune that it was you, someone I knew and could trust."

"So let me understand this. You believe my apparatus could be used to kill in-vitro; embryonic brain death? But how? I don't understand."

"The reports from the German experimenters described an electromagnetic field which could be placed over pregnant women and in 100% of the cases; the babies were born brain dead. They were attempting to understand the effects and produce it on a larger scale, but the allies liberated the camps and the primary doctors in the experiment were all executed for their crimes against humanity."

"So if they got that far, why hasn't someone repeated the experiments?"

"I believe it is a twist of fate that we arrived and got them all, at least all that had the technical expertise to continue the experiments. My travels around the world to investigate birth defects in the brain have been to keep watch on unusual occurrences. So far there have been no more than the statistical averages, no concentrations. I believe this is proof that the technology is not developed."

"In reading your research reports and papers, you are investigating the energies that could be manipulated and misused. I need you to be aware of the dangers of proceeding before you publish specific findings that might be used for this terrible purpose. You are blessed to live without direct knowledge of the atrocities that certain groups wish to befall on other groups, horrors you cannot conceive; these groups still remain on their course and still seek this and no doubt other technologies that will complete their aim. Specifically, the Aryan's and their insane notion of their supremacy. The same insanity that led to the Holocaust. Are you sure you haven't had any unusual inquiries, no strange visits?"

"Lester, I know you, I respect you, I hold you in the highest regard; but this is insanity. Do you honestly believe that any of this is more than a fantasy? I have listened and I can't believe this is you. Why didn't you confide in me earlier, given me a hint? How can I be sure this isn't some farce or dementia? You must know what I'm forced to think right now."

"Karen, I know this came suddenly to you. I can empathize with how this must appear, with no forewarning, no preceding evidence. I have to rely on our relationship, your knowledge of who I am and what I have done in my career. I know it would be difficult for you to believe I'm being followed by the NSA, but I am. I believe you are holding the technology that could be used in a very destructive way by forces you cannot conceive the depths of hatred they are ruled by. I'm here to warn you and to have your assurance that you can misdirect them if possible and to cease the research and destroy your notes if you find what I am saying is the truth."

"Lester, I refuse to do that. I have spent almost 10 years and I'm about to break through a barrier that could potentially end neural disorder without invasive drugs and surgery. That has the potential to save lives and suffering of millions of people. I can't simply throw it away on an old war story and your crazed convictions!"

Karen could not believe this request! How could he ask that, what could he possibly know, this technology is purely passive, Suzanne had said so. She was planning to make it active, but only on an individual case by case basis. It couldn't be ratcheted up beyond a single patient could it? Karen recalled Suzanne's phone call coming down the hill. Embryonic energy absorption. It was too soon to rely on, it was only a postulation, but a nagging doubt, small remained.

"Lester, I'm sorry, I just can't take all this in without some skepticism. Why don't you come to the house; meet Suzanne. She can tell you how unlikely this is. There are no uses for my technology at this point and I doubt that it is even what the Nazi's were doing. Come to the house, let's compare our experiences and I'll show you how impossible your scenario is."

Lester quietly listened and prayed the words would come to make her understand the danger and the urgency of the situation. She needed more evidence, from outside Lester's influence. What could he do?

"No, that's not a good idea. I'm known; coming to your house increases the danger you and your family might be in. But we need a way to communicate, outside of prying eyes and ears. Remember 1945, it will be a clue in future communications. I've stayed too long. We will talk again. Karen, please do not think me a schizophrenic or otherwise dysfunctional. I assure you I'm all here. Consider what I've said and take some precautions, but please understand there is no end to the resources of the

Aryan's and no limit to the lengths they will go if they believe your work to be valuable to them."

Lester stood to leave and Karen rose but Lester stopped her. "Wait five minutes before you go. I know you cannot believe this now, I hope you never see the proof of such things, but I do know and it is safer if I go, alone, in case I'm seen."

With that he was gone. Karen wondered if Lester was alright or if this was a dream. She texted JJ and said she would be home in a half an hour. JJ replied that he was waiting and curious.

36

Karen had gotten the call from Suzanne on the way down the hill to pick up Lester. She decided not to divulge the preliminary data Suzanne had mentioned; it was far too early for such speculation she thought. But after listening to Lester, and now having a moment to herself, maybe he was not delusional. If the brain is absorbing energy and that is the spark of life, then shading it would prevent life from starting at all. Maybe he was right, what to do now.

Lester warned of new people, new interest in her work, but wasn't that the nature of her business. Every time she published, she received interest from several people, some she didn't know. Lot's of graduate students, a couple of practitioners. Not always from the US, some were from Europe. EEG practitioners were few relatively; a small circle. But the part that makes his story simply unbelievable is the NSA and Nazis. Those are fantasy thriller ideas; they don't really exist in the real world. Old history, cold war history, gone today. But are they?

When in doubt, research. The mantra for any good experimenter, do not discount even the most absurd idea until it is researched. She made some notes to begin her own research but now she needed to get home to a worried husband and to look at the results in a new light.

She arrived home not noticing the new van on the street, pulled into the driveway to the door opening. JJ and Suzanne were talking in the living room and saw her headlights as she made the turn.

"What was that all about? JJ asked.

"Let's go inside." Karen said and abruptly entered the house, closing the door, locking it and peeking out the window.

Karen gave Suzanne and JJ her recollection of the conversation with Lester. When she finished, 15 minutes later, she felt exhausted and spent.

"Is he crazy?" Suzanne asked.

"I forgot to tell you that I saw Lester in Geneva last month. It totally slipped my mind when I got home. He seemed kind of dooms-dayish." He recalled his conversation to Karen and Suzanne complete with him not being at the baggage carousel in San Diego after the flight arrived.

Suzanne began reflecting on her impressions of him at their lunch meeting, one that appeared to be by chance but now seemed to be planned. "I thought he was perfectly sane, in fact, he had an air, a charisma that impressed me very much. If you were to take a vote, my experience is that he is all there."

"But why now, what's changed? Why not talk to me a year ago." Karen asked.

"That's easy, you recently sent in your report on our finding the dropouts. The grant review report, he must have read it. That's plausible. But is he right about the whole weapon thing?" JJ asked.

"After tonight I'm not sure, I'm totally exhausted. Let's go to bed, and tomorrow start researching the details he mentioned. If he was there, and if there is a record of missing documents, brain experiments in concentration camps, that shouldn't take too long to verify. Suzanne, have you thought more about the brain absorption theory you mentioned earlier tonight?"

37

The subject returned to her home at 2:04 AM. All lights in the house extinguished at 2:47 AM. All quiet. I observed the subject and husband with another woman talking in the living room for about 20 minutes before they retired. I was unable to hear anything they said. End of report.

Juergen was kicking himself for not being prepared, he did not have his directional microphone; somehow he knew that it would have been useful. But he was told to setup quickly for visual inspection, and prepare the house for complete reconnaissance as soon as possible.

He had reconnoitered the house after Karen arrived home. On the side of the house, he looked for entry points into the house and he observed some wiring from two corners of the house to what looked like 16 inch parabola dishes pointed back into the house. He recorded his observations and intended to document the house and its equipment when the opportunity presented itself. Meanwhile, he listed the necessary equipment he needed to cover every room in the house.

By 7 a.m., his replacement arrived, first announced when his cell phone vibrated. A text message asking if it was clear to change positions and relieve him. In a reply, Juergen said yes.

38

Lester had let the cat out of the bag. If Karen did as he predicted she would be researching his claims, to reconcile his sanity with his incredible story. He had taken a taxi to Poway, then another taxi back to Lakeside and back to his hidden rental car. He found two large trailers in a parking lot about a mile from the Starbucks and had parked between them. He left telltale strips of scotch tape on all the doors, trunk, and hood. After inspecting them for tampering, he felt satisfied the car was still untouched; good signs that he had not been followed to Karen's meeting.

He drove to Los Angeles, to the Hilton on Century Boulevard where he left a suitcase with business clothes in it with the bellman weeks before, rented a room, and showered. Today he had scheduled a visit to Dr. George Gordon as a favor to George, an old colleague and friend. Once in his room, he relaxed for a couple of hours, showered, changed clothes, and at about 9 a.m. he lit out for the doctor's office.

When Lester left Laguna Hospital, he had all the time in the world to dedicate to finding evidence of the weapon's development. He often traveled openly, but when he made his side trips, it was then that the NSA people would try to find him. Always wary of not revealing his secret until now, he worried for Karen now that he had really disappeared for the day. He feared it would put her at risk from the US Government as well as the neo-Nazi radicals. He was always wary of his unintentional slips, especially now that he was getting old.

He couldn't wait forever for Karen to come to her conclusions about the veracity of his story. He needed a way to assure her that he was not exaggerating or imagining this. He thought of the times George had been asked if he was in some kind of trouble. When Lester asked why, George would say he had inquiries by strangers when Lester left his office of what Lester wanted. The government sent in operatives behind Lester on many occasions asking questions of his intentions. After many of these occurrences, Lester plotted his visits to eliminate Karen Jordan's from their scrutiny.

Maybe if Lester surfaced, contacted Karen, the agents would call her and inquire what his intentions were, and that would tip Karen off about his claims. On the other hand, if Karen told them Lester was worried about a weapon so destructive he didn't even trust his own government with it, well that might get him a ticket to Leavenworth. He had to know what Karen would do, what she would say if he used the government Modus Operandi to illustrate the truth of his story.

As he entered Dr. Gordon's office, the receptionist recognizing him said, "Hello Dr. Warwick. Dr. Gordon is expecting you. I'll take you in."

George was in his office, his reading glasses low on his nose as he read the file of one of his patients with the EEG recordings showing a very sporadic Theta pattern. His door opened and Lester was shown the chair to George's smiling outstretched hand.

"Lester, always a pleasure! Linda, please bring two coffees, one with milk. Goddamn we're getting old, but you look the same as always!"

"Old and weary George. Old and weary. But never too old or weary to see old friends. How's Sarah?"

"Still ruling the roost. She says if I miss another physical, she's going to tie me up and have our kids take me by force. Still worried about me after the stroke. I keep reassuring her I'm fine, but; well you know. All's right in my world. How are you? I got a call from Geneva that you were there last month."

"Yes; I visited Scott after one of his patients had twins born with hypoxic-ischemic insults. Very rare to have both fetuses affected. He wanted a second opinion, I needed a trip to Switzerland, but it was good see Scott again."

"So still tripping around the world. I envy you, as you know. But enough of my whining. Seen anything new lately?"

"Some interesting research, the forum is loaded with talent this year. Roger Hadlyn must have had a budget infusion to get some of the names."

Suddenly Lester had the thought; I know the NSA has George's office bugged, they know I come here regularly. How can I get them to contact her without putting her in danger...?

"I'm excited about an old student of mine. Karen Jordan now, used to be Karen Hudson. She's doing some really exciting research in non-invasive modalities for controlling aberrant brain function, she calls it something like the "1945 Experiment" if I recall."

"What do you mean non-invasive? Like totally without medication or surgery?"

"Yes, George don't you read the journals? She's been conducting research for years."

"Yes, I read the journals, but I usually skip to the pragmatic, what I can get now, not what I hope for." He said short and defensively.

"George, you're taking this too personally, I didn't intend to be critical. You're one of the finest doctors I know. Tell me about your case, I was intrigued when you called."

39

Jim Wilson was the agent on duty when the irregular came out of the main decryption computer.

19:34:12z Subject 6T4334, recorded, Los Angeles, CA. "Karen Jordan", aka "Karen Hudson", phrase "1945 Experiment", Transcript to follow.

NSA monitors thousands of individual communications as well as millions of data streams for key words for the purpose of protecting the security of the United States. Out of the chatter of millions of voice conversations, data mining computers look for key words or new phrasings from monitored subjects and all recordable conversations for differences in patterns. The algorithms detect changes when specific requests are in effect, like monitoring a subject for new contact or information, and then it produces a report of "irregular activity" or irregular in the vernacular. These are sent to analysts for a determination if a follow-on investigation is warranted or if it is simply chatter. In the history of the subject's monitored behavior, he never used the words "Karen Jordan" in Gordon's office before. This tripped the pattern generator to provide an activity report.

Of course the settings of the algorithm often led to many dead ends in most cases, but the length of time NSA/CSS had monitored this subject had allowed a very strong pattern to develop. This anomaly was given a ranking of an irregular; it was out of context, new information, and in context with "1945 experiment" which was totally new. Agent Wilson created a red folder, printed the transcript, added the "Irregular Activity" stamp to the top of the folder and bagged it into a heat sealed clear plastic bag and out boxed it to the Section 9 analyst on duty. That analyst crosschecked the names, the terms 1945 Experiment and came up blank. He read the transcript and concluded the name of the experiment had non-research tones. His conclusion was this was a true irregular, and made a note to pursue this information with a field agent.

Within minutes of the spoken word, the report was on the way from an analyst who determined that a "Karen Jordan" has a research project called "1945 Experiment" and the NSA had no previous record of such a thing, uttered by the man who has been watched since 1948. This was peculiar. The analyst made his pronouncement within an hour to find out who Karen Jordan was, what brain research she was doing, correlate her with the subject, determine if this is a new threat based on the importance of the subject.

40

The day began as usual, Stephan and Jeremy sneaking into Karen and JJ's bed, this time to find them both still sleeping soundly after a long night. Karen woke when the bed moved and she heard Stephan snicker and put his finger to his lips to say, "Thhhh." They crawled over her, and nestled next to their father's feet and with a roar he was awake and the wrestling began. Karen took Jeremy into her arms and held him there while JJ dove under the covers to bring Stephan to the top of the bed.

Everyone was breathing heavily from the fracas and the laughing when Suzanne came into the room. She began laughing at the pajama clad group, but she had come to find Karen and run an idea by her that she had been up since about 4 a.m. thinking about. Suzanne and Karen went down to make breakfast and discuss her ideas while the kids were preparing for another day.

"I can tell something's on your mind, what's up?" Said Karen.

"Just more thoughts, considering Dr. Warwick's sudden appearance, his access to your grant review reports; it's not a far flung idea that he may be correct. Remember that German doctor who emailed us out of the blue? We didn't give him any information, but he was kind of pushy, wanting to know what exactly we were doing; maybe we should do some digging on him."

"So you think he may be a Nazi?"

"Until proven otherwise, it fits Dr. Warwick's profile. And he didn't contact us until the forum abstract was published.

Assuming he is a bad guy, what would he do next? I'm worried he might steal the technology. And if this could be used as a weapon, shouldn't we take some precautions? If converting this to a weapon were the problem I was working on, I don't think it is a far cry to make a net that could block the incoming rays."

"Thereby stopping the energy from getting to the womb. What's the energy doing in the womb?"

"This is what I have been thinking about since early this morning. I downloaded some data on brain development. We have been researching the measurement of magnetic fields over the head generated by electric currents in the brain. As in any electrical conductor, electric fields in the brain are accompanied by orthogonal magnetic fields. The measurement of these fields provides information about the localization of brain activity which is complementary to that provided by electroencephalography."

"We use magnetoencephalography and electroencephalography for measurement of spontaneous or evoked activity for our research. Our efforts have been to design a modality that can impress a field to de-excite aberrant brain waves, thereby correcting disorders. What we haven't considered is how that energy begins in the brain altogether. Maybe the original energy comes from the CBR."

"That would mean that life itself, the spark comes from the universe itself. That still doesn't explain the effects we are seeing."

"Think about your intuition in the parking lot, the shadow. I think the energy particles are selectively enervating brain synapse's as they develop. In vitro, the cellular development provides the physical entity which is enervated by specific particles. Maybe they begin the enervation process when the physical cells are ready for the energy."

"I thought the energy came from chemical reactions in the developing tissues?"

"I did too, but further research on that is, it is unclear where the spark comes from. The matter is oxidized, ready to have a reaction, but there has always been a hole in what kicks the first spark off. "

JJ and the boys came into the kitchen to the smell of burning bacon and two women in a deep conversation. JJ said, "I'll save it!" Laughing at the scene to keep the boys attention diverted. He ran to the stove and carefully pulled the pan off the flame and tried to salvage what he could.

"I like mine crispy" He said, looking back at Jeremy with a smile.

Jeremy said he did too and Stephan piped in, "I like it more than you," And so the breakfast began. JJ made scrambled eggs and looked quizzically at the women, knowing it must have been a deep conversation, but also knowing with the boys present, it couldn't be a complete conversation.

Karen and Suzanne were heavily considering the ramifications of how life started, how a modality could be used, if it could, and if tools they developed could be used to stop the flow of CBR. As they looked at each other, the thought that this was turning into something more important than they ever considered was becoming clear.

41

Juergen's replacement, Timothy, brought more surveillance equipment and sitting in the van, they attempted to use unidirectional mics to see if they could find some useable signal from the house until they could bug inside it.

"I think the brush is too dense to pick up anything from here. Have you tried the road behind the house?"

"Yea, I drove the perimeter and this is the only spot we have visual access from a road. We'll have to wait until there's time to get into the structure to have anything useful."

"I've received information that we are not to damage anything we find while setting up. They were very specific."

"What's so important here I wonder," Timothy said as nonchalant as possible so as not to tip Juergen off that he was questioning the orders.

Juergen; however, was much more disciplined, he snapped at the replacement, "This is not for you or me to question. We must do what we are ordered. We were not ordered to ask why. We've not worked often together, however I prefer that we stick to the protocol."

"Ok, OK. I wasn't questioning, I won't allow my curiosity to better my training again."

This had started a doubt in Juergen's mind; a bad thing in a society that relied on blind followers and could be the precursor to excommunication from the *Order*. Juergen made a mental note to report the breach of protocol in his personal log. They monitored the neighborhood with heat sensing scanners and determined there were no onlookers in any windows or joggers that could catch them unaware. Juergen slipped out of the van and into the morning air, dressed as a jogger, complete with a belt mounted pedometer as a prop, and jogged to his car two kilometers away.

He would report to his contact and round up the equipment he needed to complete his mission. He knew in three days the two women would make the trip to San Francisco; it was a question whether the father and two kids would go with them. He made a note to have the reservation checked at the hotel to see how many people they were expecting. This would determine if a couple of extra pieces of deadly equipment would be necessary.

42

George was completing his discussion of the diagnosis when Lester suggested a break for lunch. "I agree with your diagnosis," He said as they exited the building, "but the prognosis would be more promising if you were to supplement the meds with some biomagnetic therapy. There has been some research results on the Lipman theory suggesting there may be discernable effects with its use. If the patient can afford the uninsurable device, I don't believe it could hurt."

"That study has not been finalized and verified. If I recall, the Lipman study was trying to determine if electromagnetic fields can affect enzymes and cells or something along those lines. Didn't his claim include being able to tailor a waveform as a therapeutic agent? Kinda like modulating chemical structures to obtain pharmacological selectivity. He thinks the high specificity of electromagnetic signals result in the "direct targeting" of activity, without many of the side-effects common to pharmaceutical substances. Lester, it's all great in theory, but if there isn't any proven effect, the patients are simply given a dose of hope where there really isn't one."

"I recall our oath was to do no harm, not to only try the things that work in a high percentile of cases. I think the harm of the disorder is greater than the harm of trying a non-invasive therapy when the patient is aware of the studies. Perhaps the correct explanation to the patient could also provide them with an inkling of hope that could also be helpful. Just my opinion."

"As usual, you bring the issue right up front and center. Does this have anything to do with Dr. Jordan's work?"

"Not that I know of, but it is the general direction she is going. Perhaps she can answer that at the forum."

"I wish I could make it this year, the kids have planned a family cruise trip. We're celebrating our 35th anniversary."

"No need to explain, I'm happy for you both. Have a wonderful trip. Now, how about lunch?"

"The usual?"

"Of course!"

43

Hank and Megan were just awakening from another night together. They lay together like new lovers, arms entwined.

"It's so great to wake up with the warmth of a body next to mine," Said Megan. "I don't want to get up anymore. It used to be I got up because I didn't want to stay in bed alone."

"Me too; I never had trouble getting up. It's like a whole new world. But we have to; we both have to be at work at ten for the lunch shift. Wow, it's nine! We have to get up!"

"Nooo, just ten more minutes, I don't want the cold, hard air to sneak in."

"OK, but you lose the pancakes I was going to make."

"I'll trade it any day." She smiled. After a minute of reflection she turned to Hank and asked, "What is love, how can this have such a strong effect on us?"

"Philosophizing so early?" He cooed. "I know what you mean though, I feel the same way. It's so real, but it is ethereal, almost tangible but not. I don't know what it is."

Touching his chest and stroking back and forth across the top of it, she wondered how it all went together. "There must be an explanation for it. Everything happens for a reason, there must be some kind of system of rules, do this and that happens. But this, us, it's almost a faith thing; it has to be until it is defined and clear. It's almost like faith in religion, belief in the presence of a higher power."

"Well if you go that far, then what is life itself but a collection of chemicals interacting in organs that perform a function? But the brain, the most mysterious of all. It has always fascinated me. I'm constantly looking for answers to find the Tao; it is just a collection of matter but can have infinite reactions and make infinite actions. It needs power and produces thought. But what makes it start, is it a breath from God or simply chemical reactions?"

"I have faith it is the breath from God, when left without a tangible reason. I guess you choose to believe what is ingrained by your influences. Even though sometimes it is seems nonsensical."

"Inconsistent is more like it. It's not the non-secular concepts the church provides I have a big issue with. It is the rest of the edicts based on the practice I don't fathom; and the numbers of people who follow them because someone says to; because they claim to have some divine knowledge. In our own way, most of us are looking for the intangible because that is what it is, intangible. It can be defined in any way we want. From believing there is no God to dedicating your whole existence to the service to God, another infinite number of choices people make."

"In some sense I wish it were defined, but in other ways, I hope it never really is known. I think right now if love were scientifically explained, it would lesson it, make it just a biological function. I don't want it to be like breathing, I want the majesty and power of it to always be there."

"But wouldn't it still be there? If the feelings are so powerful, why would knowing how it happens lessen the experience?"

Karen thought for a minute, "Because with mystery comes imagination. If you take imagination away, then creativity and longing for it become unnecessary. You simply learn the process and make whatever you want, no unexpected revelations, no impromptu seductions, and no hope of believing there is a higher power; then it's only us."

"If I want to experience the taste of a wonderful tea, it doesn't take the good experience away to know how to mix several leaves and spices to make it. I wouldn't have to experiment every time. If I knew there wasn't a God, I wouldn't have to take the time to show the world that I have the faith;

I would spend the newfound time and energy to produce and create. I guess we'll never know though, it's too far beyond our reach anyway. But even now, if I wanted to experience love, if I had to create a mix to find you, I would do that over and over."

"I hope I never find out how it works, I only want to feel the feeling forever." And she kissed him warmly.

44

Hans checked into the Claremont Hotel, unpacked, and called his contact from the *Order*. He wanted to meet in a non-descript restaurant in Berkeley where they could discuss the upcoming strategy. After the plan was made, he left the hotel for his meeting.

"If it is not what we believe it is, we should not disturb her for the moment," Said a voice, his face hidden from view. "We cannot expose the *Order* based on a hunch Herr Schick."

"This technology is exactly what we are looking for. We must go now and obtain everything. We cannot allow it to be exposed to the public where it could be crushed or worse, duplicated by the enemy! This is the catalyst to regain our place in the world, do you not see this?"

"What power are you talking about? I know of your grandfather and his work, but as you also know, there were no surviving records. It's only your recording of the words of a dying old man. It's been more than 60 years, if this is what you believe it is we will have it soon, but protecting the *Order* must rise above a small wait. I have also read the file on the doctor, Doctor Jordan. It does not appear she is aware of the potential you describe, only her desire to make it a tool for curing disease. We'll wait until after the forum before we determine our next actions."

"But if she has some inkling of the potential, she may expose it and destroy the technology. Warwick has been to San Diego twice. We do not know if he has made contact, but if he has; what if he has shared the knowledge? We must act! I am convinced that a delay will be detrimental."

"What if he shared? He's considered a risk and a threat by his own government. He has nothing to offer her except a fantasy. No hard evidence. I expect this Jordan woman will want to keep her research alive in the hopes of helping people. We will wait until she has spoken at the forum, determine if she has made progress, and then take action. I understand your impatience, but we must be prudent. I might add Herr Doctor, in all your research; you have not come across anything. Why is that?"

"You dare to question my efforts? That is insolent! Keep your uneducated comments to yourself or you will think better of it later. I wish only to provide the rightful people their rightful place, not suffer the insolence of the uneducated. Good day."

"Herr Schick, if that's what you believe, we'll enjoy the future together."

45

Suzanne and Karen were online after JJ took the kids out for another nature run. He left a list of questions he wanted answered and an archive of data was slowly growing halfway through the research.

"So Lester was in the same unit that first entered the concentrations camps. He organized a psychological team to assist with the victims and his testimony at Nuremburg is well documented. There is a note he was relieved of duty for suspicion of taking documents. It follows his story exactly." Said Suzanne.

"What about this Hans Schick?"

"I'm searching for doctors that were tried in Nuremburg; there is a Schick, from Bremen. I don't know if we can find a relation, let's do some digging on the Bremen website for Schick's..."

"Let's say for argument that Schick's grandfather was involved and Lester is right. Why would the grandson want to carry on? Is that the neo-Nazi connection? Where do you find out who is a neo-Nazi?"

"We'll ask JJ when he gets back. Maybe he knows some international people who can help." Suzanne was still searching for war time experiments and found a vague reference to Bergen-Belsen and pregnant women involved with some testing.

"Brain death attributed to living conditions," Was all it said, Suzanne was reluctant to let Lester's story be inconsistent.

"What about the NSA? Can we find out who is being traced all the time?" Karen asked. "This is all so weird, how do you ask if someone is being spied on."

Let's email the German doctor and ask what his interest is, if it's bad, the answer should be vague, if he is real, then it should be obvious. What do you think?"

Karen thought for a moment and agreed. "I'll draft something, but keep looking for evidence of Lester's claim of being followed; maybe news stories, or some national data base for purported criminals. If he was discharged, see if it was honorable or dishonorable and why."

"I can look for the Army records, they should be public record right?" Said Suzanne.

Karen wrote a reply to Dr. Schick's email.

Dear Dr. Schick,

I am sorry for not being more forthcoming with information earlier, I have been unable to take a moment, I hope you understand. I appreciate your interest and am happy to share with you some of the results of my work. I look forward to meeting you at the forum.

You mentioned mutual work, unfortunately I do not follow all the research, maybe I have overlooked your work. What exactly are you researching and how may I assist you. Perhaps we can take a moment in Berkeley to discuss collaboration.

I look forward to meeting you,

Regards,

Karen Jordan

As she proofread it, Suzanne came over and looked over Karen's shoulder. "It's almost seems too friendly; my gut tells me this guy's a snake. But this should get him out of the closet."

"I hope we are not opening a door we don't want opened." Karen said.

After several hours and discussions of the technology, how it could be used as a weapon, and was Lester really crazy; Karen and Suzanne decided to find the boys and relax. Karen texted JJ; asked where they were, and where they should meet. They decided to meet at the Cuyamaca Restaurant where locals and the motorcycle crowd often collided. Overlooking Cuyamaca Lake, the cuisine was European roadside mixed slightly with American diner and a high tolerance for kids. It's a staple in Julian.

After making the plan to meet up with the boys, Karen was beginning to realize her intensity level and recognized the need to wind down. She had been stressed from the first dropout. "I think we need to take the rest of the day off. Let's digest what we know and start fresh later. I want to relax in the sun on the deck over at the restaurant." Karen spoke over her shoulder to Suzanne.

Rummaging through her closet for a light jacket, Suzanne agreed. As she was about to close the door, she assured herself her wireless backup drive was still blinking nestled among her suitcases. She had learned the hard way to always keep a backup and now with wireless and 300 gigabyte storage devices around, the days of floppy storage and managing them were long gone.

They went to the garage and as they were getting into the car, the phone rang. Karen didn't hear it and it went to voice mail.

46

Hans had an alert on his cell phone; he had a new email. When he finished his lunch, he went to his room and logged on. Karen Jordan, good, maybe we can get what we want after all.

> Dear Dr. Schick,
> I am sorry for not being more forthcoming with information earlier, I have been unable to take a moment, I hope you understand. I appreciate your interest and am happy to share with you some of the results of my work. I look forward to meeting you at the forum.
> You mentioned mutual work, unfortunately I do not follow all the research, maybe I have overlooked your work. What exactly are you researching and how may I assist you. Perhaps we can take a moment in Berkeley to discuss collaboration.
> I look forward to meeting you,
> Regards,
> Karen Jordan

He wrote a quick note back that he looked forward to a meeting and suggested they meet at the hotel lobby in the afternoon, on Wednesday, the day before her presentation. After sending the email, he pulled out a notebook and began writing.

He outlined his planned offer to buy the technology from her outright. From what he knew of research grants, he should be able to come up with more than enough to make it worth her while. If she would allow him to be her sponsor, that would make the most sense, no trouble, no bloodshed. He wondered how much she was backed now, certainly with his resources, he could find out. The National Socialist German Workers Party, a front for the Aryans neo Nazis in Germany, would be a willing donor if he could show the science was plausible. But that required he know more than she might be willing to share. But if he was offering to sponsor her research, wouldn't he be able to find out everything? Why does she want to collaborate? Is she fishing for information?

In all the boxes of documents he searched through, there were no clues as to how pregnant women and babies were affected. He had read of many still births, but he always assumed the camp conditions were the cause. His grossvater had said the energy in the baby's brain could be drained, that it causes what looks to the world as a fateful accident during pregnancy. He knew some of the scientists were working with energy, controlling it somehow, but the talk was to design the electronics to cover whole sections of towns. The hope for Germany and ridding her of non-Aryan races rested in furthering the concept. It was this that kept Hans looking to be the hero of the fatherland. To be able to rid the country of the pestilence of Judaism, of gypsies, of the ill select few who were not up to German ideals.

This Jordan woman had no idea she was treading on the fertile research of superior German intellect. But she might not have to know, it is not for me to speculate on the how or why it is rediscovered, it is only for me to control the force. He excitedly made a note to himself to remember to write his eulogy, including the part where he was instrumental in bringing the world order back into its proper balance.

47

Timothy watched as the car wheeled onto the street. Karen and Suzanne were in the car. This was their first chance to begin infiltrating the house. Using his heat detecting equipment, he studied the windows and street for signs of anyone and he slipped out of the van with a black cloth bag slung over his shoulder.

He walked down the driveway hiding a proximity sensor in a bush near the top of the driveway. He put an ear bug in his ear, tested the sensor, heard the tone, and proceeded to the front of the house. The front door was locked, as were the windows and he went north around the house. The trees and landscaping made it difficult to see another house, though the next door neighbor was only 60 meters away. He trained his heat detector on the neighbors and saw no evidence of human presence and he proceeded to the backyard. There were no houses behind and he was out of sight of any neighbor; he could work casually.

The back door was locked. He determined the first devices he must get operating were the glass mounted microphones with built-in 200 meter transmitters. He placed one on the bottom edge of the upper pane of the double hung window in the kitchen. He repeated the placement of another device on what looked to be the family room.

The back porch, a redwood decked patio at least 4 x 15 meters creaked as he walked down the back of the house looking for entry into the house. That would make planting the other bugs easier, but all the windows were locked. As he turned the corner on the south side of the house, he could see two neighbors' houses; there were only garage windows where he could place a microphone so he reversed his steps to try to get into the house with little telltale. His only option to get in the house was to pick the lock of the back door.

He looked around for signs of a security alarm system; there were no stickers on the windows or evidence on the doors. The sensors could have been professionally installed in the door jamb, but out here in a small town, he doubted they went to that much trouble. He gambled and pushed the door open and quickly examined the jamb for a magnetic switch; it wasn't there. He left the door open in case he needed to exit quickly should the proximity switch go off.

He went straight upstairs, identified the bedrooms, and found the office. There were five computers on large tables set up around the perimeter of the room. Plenty of places to hide his transmitting mics and he set two of them up for redundancy. He took several pictures of the room and proceeded to photograph the entire upstairs. Wires going to the master bedroom looked curious as they disappeared under the bed. He took many more pictures and set up a video transmitter in a vase in the hallway.

He went downstairs, steered clear of the windows and hid one more video transmitter in the house to have full field of view in the family room. He turned to the phone in the kitchen, found the RJ-11 jack, inserted a transmitter into the plug and plugged the wire back in. This completed the surveillance tools, all tiny and unless looking for them, virtually invisible. He backtracked out of the house, relocked the door, unlocked a window in the garage, checked for any people that might see him, and returned to the van. His next few hours were spent testing the devices. He had good signals from all of them. Now we will know everything that goes on in that house.

He made notes into the log, noting the time he spent in the house and waited for the proximity sensor or his replacement to wake him with a vibration on his phone.

48

NSA/CSS Agent Owen Tarthman let the phone ring and left a voice mail for the Jordan's. He said he was taking a survey for the San Diego Tribune and would call later. He turned to the Assistant Director and offered to call her cell phone, but the boss shook his head no.

"It's already not much of a reason to call her and ask about 'Experiment 1945' and explain how they knew Lester Warwick mentioned her on her house phone; it would be totally out of place on her cell."

"Maybe we should just pick her up; that removes any doubt about what we want and we can control her reaction."

"No, I don't want to make her suspicious, we don't have much to go on yet. We need to find her at her house where she is near the equipment. I'll call for a warrant and we need to have humint on sight. Send an FBI team to the Jordan house and when they are set up, call me back."

"I'm going to go for us; we'll need both for recon and I want to be close to this." Owen made a call, wrote the order, and left FBI Division headquarters to make the drive up to Julian. For over 12 years, Owen had been the agent responsible for Subject 6T4334, Dr. Lester Warwick. This case was a mystery to Owen in one regard; Warwick seemed to be a straight-forward guy. Most of the Subjects were legitimate criminals, didn't last as long in surveillance and off they would go to another prison typically. He had read the file and knew the doctor had stolen some papers that could be dangerous back in 1945; at least that's what the report said. Dr. Warwick's contact with the government was ignored back in the 40's, but the powers that be still wanted him watched because of the suspected content of those documents. They were purported to contain German experimentation results on some gruesome experiment and the fear was he might develop it himself, and the government didn't know what it was.

Owen had 10 of these subjects. He was constantly keeping tabs on their whereabouts; Lester was the most exciting because of his intense travel. Owen concluded it was a cover because of the times it appeared he just disappeared. When one is trained to stay on someone, Warwick's knack for disappearing was aggravating. But it was not for long periods of time. There was no real rhyme or reason, Owen figured Warwick must have known of the surveillance and got lost just to know he could; to have some time for real privacy.

It was typically Owen who made the calls to Lester when they reacquired him on the network. He was tempted to talk to him, but the job wasn't to be his friend or to contact him, it was to monitor his movements and actions, as a matter of national security. Now that Dr. Jordan was dragged into it, he wanted to know what the connection was.

49

After a leisurely meal, the kids were throwing bread to the ducks off the deck at the restaurant when JJ asked Karen if the researching uncovered anything.

"We found Lester's story about the Nazi's and his involvement are true. He has a significant body of work in the field, but nothing about energy or weapons."

"Also, he left the Army under some strange circumstances, something about missing documents. I checked the Federal register and it was discussed by the Congress, in the Federal Register there is a reference to Dr. Warwick. The details were expunged out though, so it must have been a heavy topic." Suzanne answered.

"So the doctor may be telling the truth?" JJ asked.

"He could be, but if everyone thought he was a crazy then, why should we believe otherwise now? I think he's fixated on a weapon, but that could be a residual effect from his observations of the concentration camps. It's a natural reaction to have some conspiratorial fears after seeing something like a concentration camp, suffering on such a large scale as the Holocaust."

"Karen, he didn't change his behavior toward you until we sent your last report to the grant review board. If he had these fixations, don't you think after all these years he would have displayed them earlier? I'd think a weapon he knew about would be something a normal person would want to stop." Suzanne said.

"Not necessarily. Dementia is often associated with advancing age. I hate to think of Lester succumbing to such a wasting effect, but right now that's my view. But, I'll not discount his fears altogether, if our apparatus is close to this thing, how could it be used to destroy mankind?"

"I have to design some more tests to confirm a few things, but I think it's possible the energy can be blocked prior to the absorption into the synapses. If the energy doesn't enter, there may be a period of time the synapse gets too old to take it in, like a time that is ripe for absorption, after which the un-enervated cell dies."

JJ recalled the conversation in Geneva, "When I saw Dr. Warwick at the airport in Switzerland, he talked about the effect on the world of weapon development, how we went from sticks and stones to gunpowder; and how there was always a détente, until a period where all sides acquired the same level of weaponry. His concern was this weapon would be irreversible, and could end human life."

Suzanne conjectured, "If the interferometric net was tuned to block the particles from entering a geographic area and if the particle energy is necessary to develop the brain, then he would be right. If the technology could be set up in orbit, then modulating a deflector shield toward specific geographic area, one group could sterilize another without anyone knowing how or why. Once identified, it could be difficult to stop the effect from depriving some of the energy that caused a loss of enervated brain cells. He's right actually; if the energy absorption is necessary for life. I wonder if there are linear processes, if you only receive half of the normal amount. That would have very negative effects, from loss of motor function, loss of memory; but at the end of the day, a dumbing down, almost worse than death. A future of dependants; what a scary thought."

"Are we creating the ability to do this?" Karen asked. "Suzanne, could it be done?"

"Given the key to this weapon concept is the hypothesis that energy is necessary for life, and energy can be directed, yes. Not with what we presently have around our house, but certainly it could be done, in my opinion."

Jeremy and Stephan were getting close to their limit of inattention. Karen and Suzanne each took one of the boys and found the ice cream bar inside and together they built sundaes. Karen couldn't help but see her son, with half the brain power, if this was possible, it has to be stopped.

50

Hank and Megan finished their shift at the yacht club restaurant and decided to go to the ashram for a few hours of discussion and meditation. As they walked in the main room, a discussion group was in progress, Swami Chingkoan was lecturing on the purpose of life and the quiescence of finding the path to inner peace as members of the ashram. When Hank and Megan took a mat, Master Chingkoan smiled and mentioned their arrival.

"My apologies Master, please forgive our tardiness and disturbance."

Master Chingkoan said, "Hank is one of our most committed members, his quest to find his way is far along. All of us are motivated by the same reason: the search for truth, the search for self, the search for happiness. All came to learn self discipline in order to find themselves. Many have adopted the Sivananda teachings as part of everyday life, and stay connected for a long period of time."

"The goal is to cultivate an innate sense of selfless love, sacrifice, community, and unity in diversity. The spirit of sharing and tolerance is communicated once you step in Sivananda ashram. The inner life, the conscious life, the positive values are imbibing the atmosphere of the ashram. Hank is one of our most experienced members and it is with honor that he joins us. His meditative journey has taken him to levels approaching that of the great masters."

Hank remembered today was the indoctrination class, he probably wouldn't have come had he remembered, but now that he was here he focused on the moment and allowed himself to display his inner peace. "Master Chingkoan speaks highly of one so unworthy of such high praise. It is my honor to be present and to find peace in this room."

The class continued for 45 minutes. The new members asked many questions of the Swami and some directed to Hank. After the Master's lecture was complete, the initiates went on to tour the grounds of the ashram, some to live there, some to simply become a member for peace and yoga.

Swami Chingkoan took Hank's and Megan's arms and led them into his office. "When the Tao found a way to unite you two, I was very pleased. All facets of life at the ashram reflect a central holistic teaching that all life is one and true happiness and true health engage the body, the mind, and the spirit. We need to take care of all aspects of ourselves. Yogic spiritual teachings assert the inner spirit is supreme, and governs the body and mind. We are all on the journey to self-realization. The needs of the body are not an exception; I hope your journeys have brought you together to continue and contribute to the peace and harmony within you."

"We have found ours has been a coming together with tremendous power, the power of love of another in a new and intimate way." Megan said. "It has been a merger of happiness and mutual respect."

"It has also caused much difficulty for my meditations; I find Megan a distraction that is overwhelming my thoughts."

Laughing, the swami said, "It is good to have such a distraction is it not? With distraction being the bar to quiescence, is yours not a better distraction than war or pestilence? I think not. It appears that the newness of your intimacy is the distraction, perhaps when time passes, you will find your discipline."

"You are wise, I trust that will be the case." He said. "Master, I have asked for your thoughts on my visions on several occasions, I continue to have the lights in my vision, have you completed the research you were going to start after our last meeting?"

"Ahh, yes. I have researched and found something I want

you to read; listen to this, this is a section from an article from Master Subramuniya:

> "Occasionally young aspirants burst into experience indicating a balance of intense light at a still higher rate of vibration of here and now awareness than their almost daily experience of a moon-glow inner light. *It is the dynamic vision of clairvoyantly seeing the head, and at times the body, filled with a brilliant clear light. When this intensity can be attained at will, more than often man will identify himself as the actinic force flowing through the odic externalities of the outer mind and understand it as a force of life more real and infinitely more permanent than the external mind itself.* Occasionally, through his newly unfolded extrasensory perception, he may clairaudibly hear, within, the seven sounds he previously studied in occult lore. *The sounds of the atomic structure of his nerve system and cells register as voices singing or as music of the vina, of the sita, or of the tambura.* Instruments to duplicate these sounds for the outer ears were carefully tooled by the Himalayan rishis of Asvaita Yoga thousands of years ago, including the boom and jill of the tabla, and the flute."

"This is from Master Subramuniya's article on transcendental experiences. I have a copy of the complete article here for you. I too have seen lights in my meditations, though not exactly as you describe, perhaps with this knowledge, you can go forward in your quest."

"Thank you Master, you are most kind to assist me." Hank bowed.

Megan asked, "Master, we (looking at Hank) have spoken of a quandary, I would like to hear your views." The swami nodded. "Emotion, like faith, is intangible. They are however very powerful, they can make the tangible body ill or healthy. We have asked each other the question, Does a higher power breath life force into us; the intangible soul into the body or is it simply an organism affected by tangible natural processes? What if the soul is nothing more than the result of a set of rules we don't understand? What if the beliefs of many religions were incorrect? What is faith if the basis of it is not correct?"

Swami Chingkoan studied the faces of Megan and Hank and said, "For many years, man could do nothing more than walk. No one dared to ride a horse yet someone finally challenged the paradigm. The same for the use of the wheel and the development of the train and the airplane. Man never dreamed of stepping on the moon, yet we challenged that and found it could be done. Did the faith that the back of a horse was safe or the seat on a train or plane shake the faith of earlier man? No, it did not. Once done, it was not undone. Those who profited by the faith and who relied on the support of the believers are the ones who suffered, but individual man did not. If the world were to suddenly discover it is alone, the God of their faith who breathed life into their body did not exist, the faith would transfer to another unknown power and a new group will seize the power of its mysteries. Is it not difficult to rationalize most of the world's wars given the faith nature of the passion of the warriors? Perhaps it would be a blessing for the world to not have such an excuse. But I digress, was I able to answer your question?"

"I believe you have, but I fear that the emotion love, if only a formula and the result of a tangible set of rules would not be as powerful if the mechanism of it is fully known."

"But is it not illusory, love? Are you not the same two people you were prior to meeting when intimacy crossed your paths? Would the passion not be as fierce if you knew where to look instead of randomly finding each other? I believe the forces of many positive emotions are near the pinnacle of inner realization, of oneness with self. Without a peaceful mind, love might be passed over, drowned out by a greater distraction. It might not be possible to find love until the mind is at peace, even if a formula were known."

"It still leaves one to wonder, if there is no higher power; then what does one believe in?"

"Megan, there will always be a higher power; there will always be mysteries that cannot be explained. Do you not wonder what will happen tomorrow? If God is found to not be the source of life force, would not other effects be attributed to his action or inaction? I believe we are self-directed in many ways, perhaps the knowledge of there being less godly influence on man would change people's behaviors to be more responsible."

"Master, I find I need to consider your words carefully, at first glance they are in conflict with my teachings." Said Megan, just before she bowed to the master in unison with Hank.

"Master, we have taken too much of your time, your wisdom is plentiful and helpful. We will be back soon. Thank you for the article, I look forward to reading it." Hank said.

"Is not a challenge but an invitation for learning? Seek your answers and they will come. Goodbye to you both and travel in peace." The master had a dazzling twinkle in his eye that spoke of calmness and thoughtfulness. He bowed in return and walked back to his desk.

As they left the office and went into the solarium for their meditations, Hank began reading the article. There were twenty or so people in meditation already. He finished the article and let the words wash through him, to allow the meaning of it bubble to the top of his consciousness. As he went deeper and deeper in his meditation, he began to see the lights. They seemed close; he allowed the energy to wash through his mind and tried to focus on the brightest lights. He could feel another presence, of another human being, one he didn't know. So new, so fresh. He received hints of new life, but could not identify more than subtle feelings. He found if he tried to get closer, the lights dimmed to disappearance. He returned to being an observer, not try to get closer and they returned."

It was the strongest sensation he had ever experienced when he finished his meditation. He had been meditating for two hours when Megan, who spent an hour in meditation, went to the sauna and showered. She was waiting for Hank in the garden and came in to see him just as he opened his eyes. Hank felt spent but alive in a way that had addicted him to meditation from the beginning. He took several deep cleansing breaths and stretched before standing.

He walked to Megan and wrapped his arms around her lavender smelling body.

"Mmmm, you're back. It feels so good to feel your arms around me." Her warm smile was pure joy and Hank had the same feeling of contentment.

"I'll go for a quick shower and we'll get home in time for a movie. I love you," Said Hank.

"How was your meditation?"

"It was the most intense sensation I've ever had. I had some new feelings I can't put words to. I need to think about the experience for awhile."

"I'll be waiting for you. I'll read the article while you're in the shower."

51

Lester left George and hopped a flight to San Francisco. He wondered if Karen received any indication from the government, but did not want to call her for fear they might go to more extreme measures. Take her equipment; arrest her, who knew exactly how she would be treated. He took a cab to the Claremont, got his room, and opened his laptop to check email. A quick wash of his hands and face, he logged on. He decided an indirect communication with her was possible, I wonder if a blind cc would be noticed by the authorities. What would prevent suspicion...?

He had the idea. He would write to the research board and cc her, there was a legitimate reason to contact her, and perhaps that will get her to contact him when she arrived in town.

He wrote:

> Dear Dr. Hartmann,
> I have been preparing a summary report for the research board and provide my input for funding. I will submit a detailed report for each of the grantees but I will start with Dr. Karen Jordan.
> Her work has advanced to a promising level. She will be presenting in Berkeley at this years forum, I believe we can expect an interesting presentation. I will be in attendance at the forum and strongly approve continued funding of her important work; I believe it is referred to as 1945 Experiment in her notes.
> I will be staying at the Claremont should you want to contact me.

Regards,
Dr. Lester Warwick

He sent the email and silently the bits flashed into the electrical marvel of TCP/IP at the speed of light. He received a read receipt from Hartmann almost immediately, he is in his office. Karen hasn't read hers yet, he worried.

He decided to take a walk in a city that he had many fond memories of. Down Claremont Avenue, up College Avenue past the stucco architecture of homes and businesses that were over 100 years old. He walked to the Rockridge BART Station and got on, not for any other reason than he had nothing else to do. He bought a ticket to San Francisco, and proceeded to the Embarcadero Station and walked over to the Ferry building. He watched the sailboats and tankers dancing together in the water.

He couldn't help hoping Karen would find the evidence of the truth; that she wouldn't think he had lost his mind. He was aware that someone was probably following him, but he had the thought to call her from a pay phone. On this short notice, they wouldn't be able to hear her half of the conversation. He found one and dialed her number, it went to voice mail. He left no message, simply hung up. There was no concrete reason he should be nervous for her, but he was. Two more days until the forum and her presentation, two days that could have powerful forces trying to obtain her work.

Should the goodness that can come from it, a modality that could correct mental illness be stopped? If it was at the cost of misuse by evil people, maybe it must be. But that is the way of the world, nuclear energy at the cost of a nuclear bomb, automobiles at the expense of higher carbon dioxide levels in the atmosphere. Once again he reflected about his actions and his fears. Was he right? Had he hidden the documents for a good reason? Would his life turned out different or better?

He came to the same conclusions again and again after he recalled the faces of those 114 women he freed from the barracks so many years ago. The haunting face of human death, of horror and total despair. With no one looking, a small group of people will take advantage of those they can. If nothing else, I have delayed their acquisition of it. It may become a reality, but not while I had a hand in it.

He went to The Crab Shack for some Dungeness crab with sourdough bread. Washing it down with a glass of beer, he looked across the bay at the fog rolling in from the ocean. It was almost 5 p.m. and he decided to begin the trek back to Berkeley. He tried Karen on the pay phone again, voice mail again. He hung up. It can wait until she arrives in Berkeley he assured himself.

He walked to the BART station and after getting back to Rockridge, he decided to take a cab back to the Claremont. As the cab wheeled around the entrance, he noticed Hans Schick. Schick also noticed him. Each of them knew of the other through their respective resources, but did not know each other personally. The fact that Schick was the grandson of a German doctor, one of the butchers during the war, made his spine chill. He paid the fare and walked into the hotel to go to his room. There he ran into Dr. Hadlyn.

"Dr. Warwick. Always good to see you! How have you been? Very interesting lineup this year, another of your protégé's featured."

"Dr. Jordan was my very best interns, she has made her own way though, I was fortunate to have known her, although it was likely not as a result of our acquaintance. Roger, how many Germans have signed up for the forum this year?"

"Only one. A Dr. Schick, someone I'm not familiar with. Signed up at the last minute as well."

"Did he mention what his interest is? Is he in research?"

"I didn't ask, in fact all of our communications were via email, I couldn't pick him out of a crowd. Why?"

"Just curious how far the draw for your spectacular was."

"We do have several doctors from Europe, the usual, Doctors Benevides, Smithe, and Lowery. But this has always been a North American Forum. Hey, by the way, are you again going to volunteer to remain on the research board? I have to make a presentation and wanted to report you would be."

"I believe I will, the board meeting is tomorrow, if they offer I'll accept."

"That's great news, I will pass it on and help politic for you. You have always been a strong voice and your experience is becoming hard to come by."

"Thanks Roger, I'll see you shortly."

52

As the forum attendants began filtering in from around the country, the bars and restaurants at the Claremont were filling with familiar faces in the EEG world. Vendors, purchasers, and users all mixed to discuss the work, the tools, and what had gone on since they last met. Hans sat alone, he did not know anyone. His sitting alone did not go unnoticed, his dress and demeanor appeared to others that he was also a doctor and one of the friendlier doctors couldn't help but ask him if he was attending the forum.

"Hi, I'm attending the forum, are you? Names Brad, Dr. Brad Clooseman." He stuck out his hand.

Hans looked at his hand, his eyes and said "I have no desire to meet anyone and what is your interest if I attend the forum or not?"

Taken aback, the doctor withdrew his hand and apologized for taking his time, and walked away, shaking his head. He returned to the friendlier group he was with at the bar and related the story. Heads turned with people looking at the rude person and heads shook back and forth indicating they didn't know him. Eventually the group moved on to friendlier conversation.

Hans could not fathom there are people who are simply trying to be friendly, "A waste of time," He thought to himself. "Uncouth Americans," he muttered as he finished his drink and left the bar.

He went to his room, texted a request for data and shortly after received an email. It was a report of the surveillance from the Jordan residence.

Subjects left home vacated, tools left inside. Surveillance personnel changed shift at 3 PM. Subjects have not returned as of 6:34 PM. Five phone calls, two left messages, a telemarketer and a child looking for another child. House quiet. End of report.

Hans recalled the request to find out if the whole family was going to be in Berkeley and called the front desk.

"Hello. Has Dr. Jordan checked-in please?"

"Please hold... No, Dr. Jordan doesn't arrive until tomorrow."

"Excuse me, one more thing, Is Dr. Jordan bringing her children? I have some gifts from Germany for them, and I need to know if I should package them for shipment, or if I can give them to the children in person."

"That's so nice!" Said the female voice, "It looks like the whole family is coming; there are three rooms reserved for the Jordan's."

"Thank you." Said Hans and hung up.

He texted his contact that tomorrow the house would be empty.

53

The Jordan's and Suzanne came home around seven. Karen sent the boys right to the shower; Suzanne went to the office lab to continue researching. JJ and Karen sat in the kitchen and started dinner.

JJ said, "Something simple, maybe soup and a sandwich," In response to, "What do you want for dinner?"

"To finish the conversation, I'm partially packed for tomorrow, we need to get the boys packed tonight, we should leave around eight tomorrow morning. The boys are excited to fly, just like daddy does." Karen said.

"Little do they know what a pain it is. All a matter of perspective." Said JJ.

JJ was at the cutting board slicing tomatoes when Suzanne came into the kitchen. She looked worried. When Karen was about to ask, Suzanne put her finger to her lips and walked to the windows. She pointed to a little black speck on the kitchen window and pointed to it.

"What smells so good she asked?" Still with a worried look.

She went to the phone, found the notepad, and wrote, "The house has been bugged!"

"It's Dinty Moore stew," JJ said as he and Karen read the note. "With turkey sandwiches."

"Sounds delicious," Said Suzanne as she continued to write. "I found two bugs in the office and a camera in the upstairs hall vase. I think there are more too"

Karen wrote, "Who did it?"

Jeremy screamed he needed a towel and Karen ran upstairs.

"Hold on, I'm getting one. Here you go honey. Jeremy hid behind the partly open bathroom door in modesty and Karen couldn't help but look at the vase down the corridor. Crap she caught herself, she didn't want to give away that they knew and said, "OK here it is, pretending to keep up a farce that she didn't want to see him naked. I'm not looking...," She said.

Jeremy took the towel and closed the door. "Thanks Mom."

Stephan was in his room getting his pajamas on, Karen went to check on him, and then went to her office and recalled some of the discussions during the day, felling worried now that Suzanne had found that the house was bugged. She thought she should hide some notes and equipment in case whoever bugged the house hadn't already copied it. She looked around the room, nothing had been moved, she kept meticulous notes, and the books were in the proper order on the shelves. Lester has to be right, a bugged house, someone is after the work.

She noticed right away how Suzanne had discovered the bugs. Their gear could measure the electromagnetic energies of the brain; they had to tune out all the frequencies they had detected in the house using canceling technology after they identified them. New energies which hadn't been cancelled out were on the screen, dancing waveforms of new transmitting frequencies, all in the 980MHz range. These energies were picked up by the net and were displayed on the screen. She turned off all the monitors, took her notebooks, and packed them into two storage boxes. She hoped the cameras were not positioned to see her clearing out the office.

As Jeremy came out of the bathroom, Karen said they all needed to pack for the big trip. Jeremy had dressed in the bathroom in his pajamas and was excited to help.

"We're going on a trip! I can't wait!"

"I'll get my thuitcase," And Stephan ran to get his too.

"We'll need these." Karen said as she began to pack the boys clothes. She packed them for two weeks instead of two days as the original plan had been. After the latest events, she planned to send the boys to her Mom's in Normal to stay with until she knew more. After packing, she hugged them both, worried they were really in danger and the words of Lester Warwick ran through her head. "These people know no limits and will do anything to get the weapon."

Karen and the kids went downstairs where JJ and Suzanne laid out dinner. There was silence among the adults, but the kids chatter kept a noise going while they wrote notes to each other.

"I love thith thew" Said Stephan.

"So do I, can I have your carrots." Jeremy asked.

Karen dug carrots out of her stew and put them into Jeremy's bowl as she read a note from JJ. He wrote, "I 'm going to say I'm going to the store for some shaving gear and make some calls for help."

"Whatcha writing for?" Said Jeremy.

That caught them all off guard and Suzanne said "We're, making a shopping list for the trip tomorrow. We need some trail mix for you boys on the airplane."

Karen wrote "I'm scared; I packed the boys for two weeks to go to Mom's until this is sorted out. Make the flight reservations tonight and call Mom."

JJ nodded in agreement and said, "After dinner I'll go get the road food and some shaving gear, be back in a while."

Karen tried to be nonchalant but was having difficulty, "Honey, can you also find some floppies; I need to transfer my presentation onto that format for the moderator. I forgot to get them when I was down the hill." Thinking that that would justify a trip longer than just a trip into town.

JJ wrote, "Make sure the house is locked and get the gun from the closet. Stay upstairs until I get back. Call me if anything happens. I'll call for help."

JJ felt horrible leaving the women alone, but he needed to make contact with some people who he thought could help. In his travels, he met many people from Interpol who were submitting regulations for accessing bank accounts. Hopefully he could get some help from them.

"I'll be back soon. Love you guys, be good for Mom and go to sleep early; we have to get up early in the morning. Tomorrow is our big trip!" He hugged them, much longer than he would have otherwise. He looked into Karen's eyes and tried to reassure her it was going to be OK and he squeezed her hand.

"See you soon," Said Karen.

"Who's going to help clean up dinner?" Karen said looking at the boys.

"I will," Stephen said.

"Put your plate in the sink please, Jeremy," Said Suzanne as she stood to load the dishwasher.

Suzanne looked around the kitchen for other devices and noticed the phone line had a small device plugged into the wall. She looked at Karen and looked at the wall. Karen saw it too. Someone had been in her house, she felt invaded. It was a terrible feeling.

It was then that she noticed the message light on the phone. Several calls, one message from Anthony, one of Jeremy's friends wanting a play date for next weekend and a telemarketer. Even in this circumstance, Karen thought to herself, I wish they would stop calling. Karen scanned the Caller ID and saw an unlisted call and two from the 415 area code. She pulled out the phone book and found that it was San Francisco. She wondered who would be calling her from there.

54

Owen Tarthman had been on scene for two hours, parked a block away from the house waiting for the FBI team to arrive. It had taken all day to get a team, bureaucracy works, very slowly he thought. They called his secure cell phone when they arrived in Julian at 7:23 p.m. and Owen gave them directions to his location so they could plan the stake out.

Owen left his car and got into the Ford Expedition with blackened windows surrounding an area filled with electronics gear. The innocuous antennae for finding and receiving intelligence equipment transmissions began sniffing and getting readings before they were even near the Jordan's house. Officer Bruce Johnson, the techie, said to stop so he could get a fix on the readings. While he was doing that, Owen briefed them on what he was authorized to tell them. Since the intelligence agencies were now all under one Homeland Security umbrella, the cooperation was much higher, but there were still interagency secrets. Bruce announced that by the looks of the telemetry, the transmissions were coming from the Jordan house.

"Hold on, I'm picking up multiple signals now, someone has equipment setup here. Stop here; let's see what we have...I have six microphone transmitters and two video signals." Said the techie in the back.

"Does the NSA already have onsite surveillance here?" Asked Steve Hardin, the driver.

"No" Said Owen, "Let me make a call. Can we record the devices you're seeing?"

"Already recording." Said Bruce

Owen called his boss, reporting the findings; he described

the video feed as he was watching the Jordan boys, a younger woman, and who he recognized as Karen Jordan, walking from the kitchen to the stairs, talking about getting ready for bed. It was 8:12 in the evening. From a camera, he had not authorized, already installed in the house, he was watching.

"Do we have any other operations going on at the Jordan's?" Asked Owen.

"No. So you're saying that the house is already set-up? Have the tech try to ID the equipment. Maybe we can determine who else is interested in the Jordan's. In the meantime, I'll pass this upstairs and get back to you with the next step. Do you want backup?"

"There are three of us, I don't believe we need help, I would deduce that someone is wanting to get the equipment but they will likely wait for the Jordan's to leave town before they go in."

"OK, remember you're an hour from any help. Is there a place for a chopper to land in case?"

"There is an intersection, but no fields, it's too mountainous. I'm going to recon to find out if our bad guys are still here, it could be a remote transmitter that should be easy to spot. If there's a field operator, I'll report it." And he hung up.

"Can you guys get a make on the equipment; manufacturer, serial number or anything?"

"I'll work on it. Each transmitting device has some unique methods of transmitting; I'll check our database for a fingerprint."

"OK, I'm going out to see if the perpetrators are still in the neighborhood. If I speed dial, I may need assistance, otherwise I'll check-in every 15 minutes via text message. I'm driving first to see if it is evident, then track on foot." He pulled out his map and traced his intended route; west side of the street, near the road. He drove down Jordan's street and checked for strange vehicles. "Amateurs" he thought. He saw the van and called it in to the FBI agents, "Tan Econoline van, side has a company named 'Red's Carpet Service'; that is definitely the perp." He reported the license and continued his drive by.

He parked out of sight of the van, pulled out his night vision goggles, and carefully scanned the area for someone hanging out. Standard operating procedure for the kind of surveillance the van was doing was to locate a public spot and man the equipment. The onsite person would scan for discovery using proximity sensors, heat, and motion detectors. The terrain of the neighborhood would be to Owen's advantage. He could hide behind trees and hills while he scanned the horizon for his bad guy. He was sure it was a manned op if the van was there.

As he followed the line of trees over the horizon, he saw the definite heat signature of a person in the van. He backed away and texted in, the carpet van is manned. One person, but they needed to find out who he was with, and this was as good as it gets; having a transmitting agent unaware of the resources of the FBI and NSA about to clear his pipes with the latest in surveillance monitoring. He called in and gave the director the lat/lon of the van. Shortly there would be three satellites and four cell sites coordinating a fix for any transmissions from the van. It was a matter of time to copy the data the van uttered along with who it was transmitted to.

55

Juergen logged into his notebook the time the male left. His recording devices were operating correctly. He transcribed the conversations and commented on the video. He was prepared to begin his transmission on the 9 p.m. hour. He reported:

> *Observed nothing out of the ordinary. Subjects in, male left for last minute shopping; the family is leaving in the morning. No mention of how long they will be gone. There will be a window of opportunity to remove any and all equipment tomorrow. A nighttime entry would be best. End of report.*

Each hour, on the hour, he texted his messages to another cell phone, knowing this method had been successful as long as they did not use specific times or names creating red flags of information the feds were looking for embedded in the text. He nestled in for another long night and waited for his replacement to come at about eleven. He made a note that the male should be back after two hours tops, anything longer would be out of the ordinary.

56

JJ got to town and decided it was better not to use his cell phone. He didn't know what kinds of resources were involved, but to flag anyone aware of the bugs would certainly add more danger to the situation. So he drove to the Julian Bakery and found the pay phone and called his friend in Germany, Gerhard. It's 8 p.m. here, it's 5 a.m. in Berlin; I hope Gerhard is an early riser he thought. He called the number and after six rings, the awakening voice of Gerhard Stiffler in a yawn said "Ja."

"Gerhard, this is Jeremy Jordan, sorry to call so early. I'm in need of a number you have."

"Jeremy Jordan? From America?"

"Yes, Gerhard. It's me."

"Are you aware of the time?"

"Yes, Gerhard, it is an urgent matter. I need the phone number of the delegate from Interpol. I have a situation I need help with."

"Interpol, what is it that is so important so early in the morning?"

"Gerhard, I will explain later, please can you give me the number?"

"Ja, ja, hold on...49 7113 4556 0943. Do you have it? This is not Interpol; it is the number for Herr Mueller who works there. Jeremy, what is the matter, how else can I help?"

"You have helped all you can for the moment; I assure you I will explain it later. I must call Herr Mueller right now though. Auf Wiedersehen Gerhard."

JJ quickly called Mueller and got him after four rings.

"Herr Mueller, This is Jeremy Jordan, from the ISO Banking Committee, Working Group 2, I'm sorry to call so early, but I fear my family is in danger."

"Herr Jordan, what is so important at 5 a.m. in the morning?"

"I may sound like a lunatic, but my house has been bugged and I fear it was done by a neo-Nazi group that may injure my family. I know of no-one else who I can turn to for international help, other than you."

"But why Interpol, we do not accept public inquiries, only requests from other police agencies. This is improper. You must contact your own authorities; there is nothing I can do."

"I know I'm going to sound even crazier, but I fear my own agencies as well. Otherwise I would have called them. Please Herr Mueller, I know this is unusual; however you and I have known each other for many years, I'm only asking you to send an official request to the US authorities so that it is recorded properly."

"What request? Nein, this is improper. I would be considered mad to submit such a request. What is this about, how do you suppose neo-Nazis are involved?"

JJ retold the story Lester had revealed to Karen and how they had their home bugged and there was an inquiry from a German doctor, a Dr. Hans Schick from Bremen. JJ couldn't go into much more detail because he was not the researcher, but Lester Warwick has the belief that Karen's work could be used to continue the development of a terrible weapon.

"Herr Jordan, I can do nothing, this to me rises to no level above a story, a fantasy. It is improper to use my position when there is no clear violation of the law. So far, you say your house has been bugged, but I have no concrete evidence of it. I suggest you contact you own federal agencies in this matter. Guten Tag Herr Jordan." He hung up the phone.

What a waste of time. He believed he could get help and now he felt useless. He tried to remember the name of the US representative in the FBI who attended the meeting. What was his name, Carter something, Carter Gibson. But how could he find a Carter Gibson in the FBI? He wondered. He decided to call the FBI, but had to call 411 first. Eventually he got through.

"FBI." Said the voice on the phone.

"Hi, I have no idea how to ask for help, but I believe my family and I are in danger."

"What is your name...," Said the voice on the phone.

"JJ Jordan, I mean Jeremy James Jordan...," JJ retold the story and hoped he had made the right decision. Ultimately JJ asked if he could get help to his house. They asked him for his home and cell number and instructed him to call if immediate danger was present.

"The report has been taken and I assure you it will get passed to a field office and an agent will contact you. I have noted you feel your cell may be compromised, so do not use it. We will be in touch with you." The phone line disconnected.

"Great, don't call us, we'll call you," He muttered looking at the dead phone in his hand. That took another 20 minutes. He needed to call the airline to get the boys tickets and talk to Karen's mom. He retold the sketchy story. Very alarmed at the hour and the story, Judy Hudson listened to JJ and said to send the kids but keep yourselves safe. I will call the FBI from this end too, in case they think you're a nut or something. JJ thanked her and told her not to worry, it is probably nothing, but having your house bugged was certainly not something you *don't* worry about. JJ said he had to go back to the house.

"Be careful JJ."

"I am Mom; if we can get through tonight without them suspecting we know about the bugging devices, I think we'll be fine."

JJ was worried. Should he call the local police, wouldn't that tip off the people who were bugging the house? No, he shouldn't call anyone else, he just wanted the night to be over and get everyone out in the morning. We could leave now and stay at a hotel near the airport, but would that tip off the bad guys? He decided to make that decision with Karen and Suzanne when he got home.

57

Hank and Megan had gone to the movies. Somewhat out of the norm, but they went to see the Last Samurai and were commenting on the details of the movies philosophy. Honor in killing, self-sacrifice if found dishonoring the system. They both were questioning not the honor of the system, but the lengths these people went to support it.

"But is it any different today? Just different weapons," suggested Hank.

"I don't see any honor in any killing, even to kill those who kill others. Just take away their freedom forever." Megan blurted.

"So the death penalty is not something I should be in favor of, ehh?"

"No you shouldn't and let's just walk." Another misty night on University Avenue. They walked down Telegraph and stopped for a tea at the bookstore before heading in for the night.

"Who was that author the swami introduced you to? Master Subramuniya? I wonder if he has other writings." Megan said.

"That's him, over here. I thought I had read a lot, there just never seems to be enough time to read even a fragment on a subject. He has over ten books on meditation."

"But you really didn't read about meditation, you had the real thing, with many masters in person. Maybe that's why you are so good at it while people like me had to read and struggle to find how to meditate. You're so advanced; maybe it's time you wrote a book."

"You're always flattering me, I almost feel embarrassed. I admit I have gone far, but in no time you will be able to do the same thing. I just want to find an answer to why I see the lights."

"Subramuniya talked about the physical nervous system, seeing and hearing the actinic voices. It sounds like what you are experiencing."

"But more than that now, after the last session at the ashram, I associate it with new life, of the soul being born. It sounds funny to say. I hope there are more answers in his other writings."

"I'm still wondering about the Masters talk today. What if there weren't a God? What do you think he meant?" Megan asked.

"I believe he was saying we all are independent, that we should consider our existence without reliance on unseen faiths or believing in things that others want us to believe. If we experience life with the point of view that our actions are purely our own choice, then the reactions we get from others are a direct result. If bad things happen, we should not attribute those results to some unforeseen force, we should attribute them to our own karma. My search has been to find the karma I possess and act to maintain a positive mind."

"What about God. I believe it would shake the faith of many if there weren't a God, if someone found a means to show the soul is simply biological mechanisms. How can we continue to believe in anything if there isn't a higher force, a higher power? Think of the millions who dedicate their lives to worship and faith to that power. Where would they turn?"

"If everyone were trying to find their Tao and keep their own karma positive, would they rely on a higher power to find peace and happiness? I think not. It would be confusing at first, but like riding a horse, they would learn the new ways and go on. If such a thing were possible, knowing that God does not exist, then as the Master said, the losers would be the churches and organizers of the faiths; but in the end the people would find another path to follow. I can clearly see how the Master can simplify such a circumstance. Would you live life differently?

Are you not seeking your Tao? If your Tao was to know you need to be self-sufficient, would you not accept it?"

"My faith would be shaken, I would need to have something to bridge that change. It would change how I looked at life and the world. Do you believe in God?"

"I know that I don't know. My search for meaning is not based on any premise that one exists or not, I have no preconceptions. I meditate to seek the truth and accept what my reflections and visions tell me. I have never seen the finger of God touch me or anyone; I see and feel that my actions in this life are under my control. Perhaps we are simply leaves in a stream. I believe we should fashion a rudder and a sail and have some control on where in the stream we can stop and some control to miss the dangerous rapids. Does God steer my leaf, I think not; more if there is a God, he gave me a brain and intelligence to steer for myself."

"It scares me, I've always looked at life believing, having a faith; that God will be there when I need Him. I cannot simply turn that off, I would have to consider my trust in Priests and Ministers I have known all my life. It would mean they are truly unaware and their dedications were for nothing. Hank, there are so many people who believe in the concept of God, what would fill that void?"

"You were the one who theorized the scenario; perhaps it is your destiny to answer that question. You must have had some questions, to join the ashram, to find alternatives to the voids in organized religion. If you take this time to reflect and meditate as though there were no God, to only reflect on what is tangible and what you can see is real, you may answer your own question."

58

The *Order* received the 9 p.m. report from their surveillance team at the Jordan's house. "Tomorrow we will negotiate with the woman or take the equipment outright. Herr Schick, if we acquire the equipment, can you determine from the notes yourself how to continue the research, or will you require assistance in this?" Asked the voice on Han's cell phone.

"I can decipher it, I have the highest belief we can continue without anyone else being involved. I believe we should keep the knowledge to ourselves."

"So the plan is to talk to the woman on Wednesday afternoon, and based on her conversation, we will send in a team to acquire the equipment or acquire it by an offer of support. Ja?"

"Ja."

"We will not contact each other until after your meeting. Contact me in the usual way then."

59

At 9 p.m. the NSA intercepted 17 cellular signals from the four cell sights and had reconstructed them for review. The analyst instantly determined that one was not the ordinary.

Observed nothing out of the ordinary. Subjects in, male left for last minute shopping; the family is leaving in the morning. No mention of how long they will be gone. There will be a window of opportunity to remove any and all equipment tomorrow. A nighttime entry would be best. End of report.

The phone number that is was texted to was registered to a German national in Oakland CA, a Roger Deichmann. A simple search of Deichmann revealed he was a member of the *Order*, a neo-Nazi group known to have been active for years. He was acquitted on several assault and battery and conspiracy charges which were dropped for lack of evidence. The *Order* was one of the highest priorities within US intelligence communities. High on the list, this was a good day.

The analyst completed the report and contacted his supervisor who added it to the day's find of other collected data and reported it to the assistant director of the North American Subversive Task Force, Western Region. He made a note of the interest by an NSA agent named Owen Tarthman with a recommendation that he be contacted immediately.

The STF assistant director made the call personally to the NSA who forwarded the call and shortly afterwards Owen hung up his phone. He told his FBI surveillance team they were dealing with the *Order* and this was certainly a curious turn of events. Owen called his boss and reported the information, requesting orders.

By 10 p.m., Owen was told the FBI had been contacted by a Jeremy James Jordan who reported he believed his house was bugged and he was afraid for the safety of his family and didn't know what to do. "Tough cookie," thought Owen. It is probably eating at him that he is vulnerable and doesn't know who he is up against. He needed to contact him without tipping off the neo-Nazi Aryans, so he considered texting him. He ran it by his boss and after a few minutes, the orders were to do nothing unless the other group made a move on the house.

Owen and the agents reread the text message. The plan was to remove the equipment. What equipment? Owen knew more than the FBI but was not authorized to reveal the Lester Warwick connection. Owen had to consider alone what this latest development meant in the big picture. He recalled that Warwick was a subject because he had stolen German documents he claimed explained a weapon of tremendous power and he didn't even trust the US Government to be able to resist using it. Maybe this Lester Warwick was on to something. There is little reason for Nazi's to be surveilling a house in Julian on a Tuesday night, talking of taking some equipment. What Owen needed to know was what Dr. Jordan was working on. But he was here and there was a present danger to the Jordan family, he did not want to leave. He went to the back of the Expedition and asked if they were connected to the web.

"Not at the moment, but we can." Said the agent.

"OK, here you go, not very fast, only 56k, but it is working."

"Thanks. Owen did a Google search for Dr. Karen Jordan and discovered she was a psychologist turned researcher. He went to the website of the forum as it was the second hit on the list. He read her abstract. It still didn't click. He spent another few minutes reading what was publicly available on Karen

Jordan, noting she was not a criminal, not aligned with any subversive groups, was in private practice, mother, yada, yada.

He concluded he had one other source he could ask, but it would probably not be within the protocol of the NSA.

60

JJ returned home and wrote to Karen and Suzanne that he had contacted his Interpol contact with no luck and he finally called the FBI. He also had called the airline and Karen's mom and she was expecting the kid's. Suzanne wrote she had found an unlocked window in the garage and assured that everything was locked now.

They made small talk for the benefit of the bugs, but deep down he was very nervous. That was also true for Suzanne and especially Karen. When JJ got home, Karen had seen the headlights, stole downstairs with the gun in the pocket of her robe, and was relieved JJ was the one walking up the driveway.

The van came to life when the proximity sensor indicated there was something in the driveway. Juergen noted the time and expected the recording devices to begin and they did. Just a normal welcome. No other talk about any experiment. In 20 minutes the bedrooms lights went out and the subjects retired to bed.

He composed his 11 p.m. hourly report:

Male subject returned to house at 10:27 PM. No significant discussion. All lights went out at 10:58 PM. Replacement due to arrive within the hour. End of report.

Juergen texted the message and sent it to his contact. He was expecting his replacement any minute.

Almost immediately, the team got a message; Owen and the FBI were now on the alert for a replacement. They had to scatter, if the replacement used conventional tools and methods, they would be discovered. This meant the house would be unwatched for a short time. Owen made the decision to hike to the back of the house to keep a watch and text if problems began. He took his night vision goggles or nvg's and disappeared into the night.

The Expedition started, reversed into a driveway, and slowly drove down the road away from the Jordan's house. They passed a late night jogger going up the road in the opposite direction.

Owen was in the brush when he got a text from one of the FBI agents, jogger in a non-reflective suit running your way. He quickly scanned the road and saw the figure about half a click away. He laid flat in a depression to eliminate any heat signature and watched as the jogger disappeared over the top of the hill, toward the van. Quietly as he could, he moved up the hill, in heavy brush. He had to take the nvg's off to see his way. Each time it took a short time for his eyes to readjust. He got into a position where he could just peek over a rock and see the van.

The jogger stopped and bent over as though he was tired. His hand disappeared into his jacket and pulled his nvg's out. Quickly Owen turned his off so the infrared light used in the goggles wouldn't give his location and technology away. He lowered himself into a position to be unseen and waited for the replacement to see his way clear to enter the van. This was a gamble he really didn't want to take, but it had to be done. If he was seen, it could blow the opportunity to catch them in the act and put his and the Jordan's lives in jeopardy. After 15 minutes, he looked around the rock and saw no-one near the van. He donned his nvg's and observed two heat forms inside the van. Relief waved through his body. He traced a path out of the line of sight through the backyards of the neighbors into the Jordan's backyard. He established a position behind a large oak tree, hiding himself while he texted the agents his situation and position.

In the van, Juergen and his replacement, Timothy, were updating each other.

"We've been ordered to observe, nothing more until

Wednesday at 8:00 p.m. We are not to raise any suspicion until then."

"Very good. Tomorrow will be an easy day. The log is up to the minute. I will return at 11 a.m. tomorrow, same protocol." Juergen checked for heat signatures outside and slipped out of the van. He jogged down the street and past the empty Expedition parked with both agents positioned in the brush down the street. With their nvg's, they both followed the jogger as he traced a path almost two kilometers away. They copied the license plate and photographed the jogger for identification.

61

JJ and Karen lay in bed and passed a legal pad back and forth, having a difficult conversation made worse by having to write it.

"If it is Nazi's, there must be something to Lester's claim. When we get to the airport, I need to call him and let him know he is in danger too."

"We haven't heard from the FBI, why not? If there were truly a danger, I'd have expected they would call, at least to say not to worry!"

"When the kids are on a plane to Chicago, we need to decide if someone should stay behind to watch the equipment. Nazi's or not, I don't want to lose the equipment. Can you stay at the house? Get Jonesy and your old frat crowd and make a party out of it?"

JJ shook his head in affirmation. "So you get to Berkeley as soon as you can and get together with Lester. Suzanne will have to fly with the kids to Mom's house. Keep in touch with me by cell. If he has spent this many years involved with this cloak and dagger stuff, he should know what the next move will be."

"I can't believe this is happening. We were so close to a breakthrough, what could I have done different?" Karen wrote.

"You are the smartest woman in the world. Nothing about this is something you should blame yourself for. And don't give up the quest. It may still be possible to salvage the modality without having the brain energy revealed."

"But now I have a choice to make, if blocking the normal brain development can be done in a widespread way, there are those who probably would use it. I wish I had chosen pediatrics instead of psychiatry."

"Once the kids are safe, we'll get to bottom of it. I hope the cavalry are coming though."

"Me too!"

"Goodnight baby," Karen said out loud and kissed JJ.

"Sweet dreams beautiful." Said JJ before he sat up, armed with his 9 mm Glock, waiting for any creak or a noise to investigate.

62

Owen called in for a report on the location of Lester Warwick.

Sam read off the log, "He checked into the Claremont Hotel in Berkeley, room 578, took a stroll to San Francisco, met no-one, made two calls from a pay phone, ate alone, and returned to the hotel. Presently he is alone in his room."

Owen knew the methods of the NSA and the resources at hand when it came to following a subject. He had thought of calling him, to fill in the blanks. It may risk his career, but it may save the lives of innocent people. He decided to make the call.

He dialed the Claremont Hotel, asked for room 578 and it began to ring.

Lester was asleep when the phone rang. The phone rang a second time, he was now almost fully awake, rubbed his eyes, and stretched his arms wide. He remembered where he was again, why he was here. By the third ring he was beginning to go through the realization that he had given some information to a couple of people, but he knew none of those people would call him at midnight, unless it was Karen.

"Hello," Lester said.

"Listen, but do not reply. Call this number; 858-555-4323 from a safe location and ask for Robert Wagner." The line went dead.

Lester did not recognize the voice. He was disappointed it wasn't Karen, but dressed and went to the lobby and asked the clerk where he could find a snack.

"I'm sorry sir, room service is closed."

"I know, I'm up for a walk and am looking for a small store of sorts."

"There's a convenience store on Ashby Avenue, but it's a long walk sir."

"That's Ok, I'm old, but I can walk. Which way is it to the store?"

"Well it's out the front door to Claremont Drive, turn right toward Ashby, and then left about a block on the left. Are you sure I can't call you a taxi, sir?"

"Thank you son, but I enjoy the night air."

Lester began the walk, wondering who it was who called. It would have been the first time anyone in government had called him. He went through the last few days; he had deliberately left a message for the NSA to tie Karen in, but only to help independently verify his story. Had his efforts caused her even more trouble? Why would they call me if Karen were in trouble? He didn't make any sense of it all the way to the convenience store on Ashby. There, near the store was the pay phone he recalled would be there. He called the number and a chipper voice said, "Hello, Can I help you?"

"Robert Wagner please," Said Lester.

"One moment please..." The phone line clicked with the sound of the line being forwarded. It clicked for almost 30 seconds before a voice came on the line.

Owen said, "Dr. Warwick?"

"Yes, who am I speaking with?"

"Owen Tarthman, NSA. I'm violating several laws, but I have a situation and I need direct information from you."

"How am I to know if you are NSA or anyone else? I'm not inclined to speak anymore."

"Doctor, I'm going by memory, but today at approximately 10 a.m., you were in a friend's office in LA and you mentioned something called the 1945 Experiment. Who else might have had the opportunity to know about that conversation besides us?"

Lester considered the possibility of George being involved with the *Order*. No, he's a veteran, his wife is Jewish, and he converted when he married her. "That is still not enough, how else can you prove you are who you say?"

"I called you at the Santee Inn last month when you had one of your rabbit outings. It took all night to reacquire you. One day I would like to know why you try to fall off the system, but not right now. There are some life and death situations I'm hoping we can end with little or no damage."

"Agent Tarthman, I need one more bit of confirmation. Are you in a directory of publicly available employee lists I can confirm? As you say, this is a life and death situation and I cannot risk anything without concrete confirmation."

"I understand. Please wait where you are. Sam, Code Sierra Oscar Sierra Alpha." The phone line clicked and a voice entered the conversation.

"Yeah boss?"

Sam, can you pick Dr. Warwick up and provide any credentials it takes to satisfy him that we are with the National Security Agency of the United States. This is a very high priority."

"I need an authorization for that."

"Call and get it."

"You got it boss." The line clicked. "Dr. Warwick, I have to have your confidence and I believe you know why. Please wait there for a car. You will be asked if you have a cigarette lighter. Accept the ride and I will call you again."

"I will not say another word unless I'm assured of the veracity of your story."

"I respect that doctor. I would do the same in your position."

63

Hans had gotten the message after the last conversation with the local contact. The comment that he had not discovered the energy himself was insulting and it did not sit well with him. For years he searched thousands of documents for Blood & Honour, the Aryan organization he had secretly joined so many years ago, with his grossvater's vague direction and little resources because his results were so scant. He studied everything about magneto-electroencephalography, but was no closer to finding anything his grossvater spoke of. And now, I am ordered to wait like a child, with the technology available with no guard on it. He wanted the technology. He didn't want to wait another day. His patience was at the limit.

He was frustrated and was now going to be in the dark for a day while these Americans, with conquering designs and calling themselves Aryans, stalled to protect their own safety. The more he thought, the more critical he became. They were cowards. He considered he had a day to fly down to San Diego and return. But they were watching; I would not able to get into the house without them knowing.

His only contact in America was the voice he never met. He didn't understand why they couldn't meet in person, again the cowardice. Hans was becoming infuriated that he had been led to believe they would help him and the B&H when they were only waiting, to protect their own hides. A true German doesn't need to hide his intentions from the world. He acts. He had acted, for years; he completed all the tasks set out for him.

His goal was to find the answer and he accomplished his goal. Why can't they simply go and get it?

He tossed and turned then he paced his room. What else can be done...? There is always a way.

64

Lester was picked up by a dark blue sedan with the appropriate password given. This is still not proof; the Nazi's have equal resources. He sat silently while Sam took him to a building without any visible markings of who owned or leased it. They drove into a parking garage. They entered through a key locked door and walked to another door with a keycard security card required. They stepped into this room not fully lit, rushing down another hall to a vestibule where a retinal scan and thumbprint security system waited. Sam completed the security procedure and the door opened to two armed Marines who accepted Sam's outstretched handgun. They asked Lester if he had a weapon.

"No, I have no weapons."

"Would you mind if we check your person for contraband?" Said the Marine before moving towards him. It appeared the question was a formality.

Lester lifted his arms as though going through security at the airport.

"Thank you sir, sorry for the trouble. He's clean sir." He said to Sam.

From the vestibule, they approached another security system and Sam slid his card, entered a code into the keypad and the door whisked open. It was a room with several computer monitors, evidently on but unless sitting in front of them, Lester could not read a word on any screen. Sam went to a desk, one of ten desks, picked up the cordless phone, and called a number.

"He's in. I sure hope you're right about this." He handed the phone to Lester.

Lester spoke into the phone "This still doesn't prove you are NSA or even with the US Government. How are you going to demonstrate it?"

"Doctor, you are very adept with computers. At the desk in front of you is a computer. Please sit and prove you can get access to the internet."

Lester surfed the net looking for random sites and assured himself he was in fact on the net. "OK, I can see the internet."

"Now search for any high ranking person in the government you would know personally."

Lester thought for a moment and recalled Dr. Joseph Fitzsimmons; he was the Under Secretary of Health and Welfare. Lester knew him personally. He searched for the doctor and found he was still in government, living in Washington, DC. He was still the Under Secretary. "OK, I have located someone, now what."

Sam said "I have a call into him right now. A few moments passed and a sleepy voice answered the early morning phone call. Sam said, "I'm sorry to bother you at this hour sir; there is a National Security issue requiring your attention. I'm Sam Valides with the NSA. We have someone who knows you, requires you to confirm our veracity, and would like to speak to you. Authorization NSACSS1. Dr. Lester Warwick."

Sam said "Doctor," as he handed the phone to Lester. Lester gave the first phone back to Sam.

"Hello Joe?" Said Lester.

"Lester? What are you doing tied in with the NSA? This is all very unusual."

"Joe, what did we talk about when we were in Seoul for the International Mental Health Forum the night Jerry spoke?"

"You're not Lester. We were in Busan, not Seoul."

"You're right Joe, it was Busan. This is Lester, I was just checking. What did we talk about there?"

"George's heart attack."

Lester handed the phone back to Sam. "OK, I'm convinced."

"Dr. Fitzsimmons, thank you for your help sir." Lester

wondered if there were some sort of understanding with all high ranking officials that if NSA calls, don't question; but it was just a fleeting thought. "OK, Owen here is the doctor." He handed the phone back to Lester.

When he heard the phone transfer over to the doctor's hands, "OK doctor, I kept my part of the bargain. Will you help me?" Said Owen.

"What do you want?" Asked Lester.

Owen briefed him on the German's surveillance at the Jordan's home. He explained the communications the NSA intercepted and explained they were delivered to *Order*. "Do you know what the *Order* is doctor?"

"Yes, I am all too familiar. An American sect of the neo Nazi movement."

"The Jordan's know their house is bugged. They called the FBI earlier for help. What I need to know is what the Nazi's want. I know your history and the claims you've made about a weapon developed in Germany, I'm concluding Dr. Jordan has stumbled onto something that has them interested in the same thing. Am I right?"

"Yes, you are, The *Blood and Honour* group, an Aryan sect in Germany, has an operative from Germany at the forum here in Berkeley, so they are aware of its importance, but I have no concrete evidence of their plans."

"We know they know the Jordan family is coming to Berkeley tomorrow and their latest intercept from the surveillance says no action will be taken until Wednesday at eight. What is happening on Wednesday?"

"I don't know. It's the day before Karen is speaking at the forum, but beyond that, I don't know. Maybe they want to be sure no-one will be home, and one way is to assume that they will all be in attendance at her presentation."

"Dr. Warwick, what is your intention in all of this? I have to know."

"My intention is as it has always has been; to prevent this technology from ever seeing the light of day."

"But what is it? What does it do?"

"I don't see any more reason now to tell the government than I did many years ago."

"It appears to me there is a family's life on the line at present. In order to prevent their injury and damage to anyone else I would send in to protect them, I need to know the nature of the weapon. I cannot put my people in jeopardy and I won't if I can't get the basics from you."

"What is your name sir?" Asked Lester.

"Owen Tarthman."

"Mr. Tarthman, I have the same concerns now as I did many years ago, however I can assure you no-one will be in danger of any kind relative to what Dr. Jordan is researching."

"Dr. Warwick, that is not good enough. I need to know the exact nature of the weapon."

Lester considered his position for a moment. Though Tarthman was being careful and had good reason to ask for the information, Lester still didn't want any government to have the benefit of the potential research. But if there are Nazis outside of Karen's home, well, it was time to put up or shut up. He took a minute to think.

"Doctor Warwick?" Owen said.

"Yes, yes. The reports from so many years ago described a method to blanket whole cities with a shield that causes natal brain death. It is a weapon that makes no sound and causes no other injury. It was the master plan to secretly blanket whole races for several generations; the Aryans would then be in the powerful position of destroying everyone."

"And the proof you have that it is possible is what?"

"Why, isn't the proof that you have been following me around for 60 years?!? It's in the documented experiments confirming the death of over 2,000 newborns, in a two year period. The German soldiers raped the women prisoners to impregnate them and then placed them in a room with some kind of shielding and every baby was born brain dead. I tried to warn several superiors but I was called naïve and too young, that I worried for nothing. I made a decision, given the evidence I had then, no one could be trusted with the technology given the degree of success of their horrid experiments. We were fanatical then, even my own government. I decided to keep the secret and watch for evidence of this kind of weapon. I have dedicated my occupation and freedom to discover and stop it if I could."

"Karen Jordan is a bright, caring, and empathetic researcher who has come close to discovering what that energy does to natal brain development. Suzanne, her assistant is a very capable scientist in her own right, according to Karen. By now I'm sure they have deduced something, if you know how the energy goes in; you can discover how to stop it. But now I must prevent harm to the Jordan's and once again I'm forced to trust the US Government, where I tried to, so many years ago and was called a nut, a crackpot. I hope this time the government will see fit to prevent this weapon from being deployed and not be so arrogant as to use it themselves. To that end, I want you to know and those also listening, I have secreted what I know and should I disappear, the secrets will be revealed to many esteemed people, where the world can be made aware and not allow us to destroy ourselves."

Lester was breathing heavy, he was tired of holding this secret, and now after telling it, he felt a strange relief he didn't expect.

Owen was speechless as he listened. "Doctor, I don't know what to say, I...," He thought for a minute, recover, be professional, what is the need now..."Doctor your information is invaluable. I will see to the protection of the Jordan's, and we will get to the technology before anyone else. Can I talk to Sam?"

Lester handed the phone to Sam. Sam crossed the room and talked for a minute. He shook his head in agreement and hung up. He walked over to Lester and asked, "Are you ready to go back to the hotel doctor?"

"Yes, if you don't mind. Thank you."

Owen texted the FBI agents in the Expedition and asked if he could rejoin them in the vehicle. They gave the all clear and he hiked through the back yards to remain clear of the van and any sensors in the area. He found his rock and looked at the van with his nvg's. He saw a reclined person next to a couple of heat generated signatures from the monitoring equipment.

"We saw another jogger, probably the relieved perp and we were able to follow him, got his picture, and made the car. The report was sent in a while ago, no word yet if there is a positive ID."

"That's great work guys. If you ever want to change to the better department, give me a call." Owen smiled and said. He briefed the agents, made a few phone calls for help, and updated the situation to his director.

"By the way, we finished the evaluation of the transmission signatures, German made. Heller Manufacturing, probably imports. Can be bought over the internet, but the power output of these have been increased; only a very sophisticated user could amplify these devices."

"What's our next move Owen?" Asked the driver.

65

JJ stayed awake the whole night, he was yawning when Karen awoke at 6:30 a.m. She took the legal pad and asked if he had any sleep, JJ shook his head no.

"Ahh, it's too early" She said. "Why did we take such an early flight? Remind me?"

JJ said, "Because we wanted to get there with time to walk the campus, remember, your idea."

"Why didn't you argue against it harder!" She joked and went into the bathroom. She shuddered to think there might be a mic or worse a camera in the room with her.

As she finished brushing her teeth, she could hear JJ and Suzanne downstairs getting coffee going.

She changed clothes, and went in to the boy's room and checked that they were OK. Still sound asleep, their innocent faces poking out from their blankets always warmed her heart. Life, that I had a hand in making. It was a fundamental force of nature, to create life. She thought of the horrible effects of preventing life at its inception. She shook her head to clear the thought of it.

The coffee smell was wafting upstairs and Karen followed the aroma down to the source. With worried looks and smiley comments for the eavesdroppers, they passed notes and tried to make morning conversation.

Karen said, "I know we're forgetting something, I always do when I go to these things."

Suzanne wrote she checked the number of devices and found two broadcasting video information and actually gotten the signal to display on her laptop. They were located in the hall upstairs, one downstairs, into the living room, but none in the kitchen. Here there were just microphones. "Karen, you always say that and it turns out its not that you forgot something, you forgot where in the suitcase you packed it. You are so organized."

Karen wrote, "Do you think we should bring some of the gear?" And said, "We should wake the boys about seven, get them into the shower, and have a big breakfast."

"I'll start making the waffles." JJ offered still in his pajamas. He wrote, "Are there key pieces that won't be too noticeable, we could send to your mom's? That way no matter what happens, they can't get it?"

"I'll heat the syrup." Suzanne said as she finished writing, "That's not a simple thing. There are not any key things, the real knowledge is in the hard disks and notes. We would have to delete everything and just keep a copy on one hard drive. Then the physical stuff is essentially useless without the software."

"Agreed" Wrote Karen. "Do you still have that wireless drive in the closet?" Suzanne mouthed a yes. "I'll start deleting the data on the desktops. The notes we will simply have to risk hiding. We can't get a couple of boxes down the hall with a camera recording it."

Karen said, "I have some more notes to get printed for the forum. I'll be upstairs."

"Don't forget to get the PowerPoint onto the floppies." JJ said as she left the kitchen.

Karen went into the office and verified the 300 gig drive had all the files and began deleting the files from her main research computer. She printed out a recipe for stuffed turkey Hawaiian just to make her story sound true. When the files were deleted, she ran a disk wiper program three times. Suzanne had told her of a guy she knew that worked in the government who had to design a way to delete files from a hard disk that left no traces. It seems the latent dipoles could be read after several read/write cycles. Newer wipe programs were supposedly better than average but at this point it was the best she could do.

She went to the other computers and repeated the deletion

and wiping. She looked at the boxes of data and decided maybe she could bring them all. She would simply announce these were the handouts for her presentation.

At 7 a.m. the boys got up, excited about the trip. They ran downstairs screaming that today they were going flying. “I can’t wait to go flying. When are we going?” Asked Jeremy.

“After breakfast and a shower.” Said Suzanne.

“Why a thower? I hate thowerth.”

“Cause you stink, that’s why. Ha-ha,” Laughed Jeremy.

“Be nice to your brother,” Said Karen coming into the kitchen. “She wrote to JJ, I think we can get my notes out of the house by saying they are handouts. What do you think,” JJ gave the thumbs up and Suzanne read the pad and smiled.

“What’s that Mamma?” Jeremy said pointing to the pad.

Somewhat taken back, Karen stammered out, “It’s a list of things Mamma has to bring with her to her meeting, honey. How’s breakfast?”

“OK, what’s on the list? Do you have the trail mix on it?”

“Yes sweetie, that’s on the list.” She smiled at Jeremy but the nervous look she gave to JJ almost melted him. This is not right, to live this way, not for a minute, he thought to himself. He committed to ending it in any way he could.

He announced he was going to take a shower, “Cause I like to, and I bet the first boy to start a shower with me, I can beat him.”

Always up to any competition, Jeremy said, “Can not. I’ll beat you any day.”

Stephan, never one to be left out, said he would too beat him and the three raced upstairs.

Suzanne said, “Karen, we don’t want to forget the handouts; they’re in the office, the two boxes. Dr. Hadlyn would be very disappointed if we forgot those.”

“I do have those on my list.” Karen said.

66

At 7:45 a.m. Owen Tarthman was donning a cap, driving a limo, and pulling into the Jordan's driveway. He tripped the proximity sensor and Timothy noted the license plate number, photographed the car, and made an entry into the log. In his hourly report, he would send the data for verification of the car.

Owen rang the doorbell. Karen froze in her tracks. Who would be at their house at this hour? JJ was still dressing and probably didn't hear the doorbell.

Karen peered through the window at a normal looking man and a limo behind him. We didn't order a limo; there is some kind of mistake. Leaving the chain on the door, she opened the door and asked, "Who is it?" Her eye just peeking around the door.

"Ocean Beach Limo, ma'am. Ordered last night for 7:50 a.m. pick-up. I like to be prompt; I'll be waiting outside when you're ready. If you'd like, I can help load your luggage."

"Hold on a minute," She said and closed the door.

Suzanne was out of sight of the camera and asked, "Who is it"

Karen answered "A limo." She wrote, "I didn't know JJ ordered a limo; I have to check with him." She went upstairs and wrote down the question, "Did you order a limo last night?"

JJ wrote, "A limo, I didn't order a limo last night."

"Good guy or bad guy?" She asked.

"Well the bad guys would just take what they wanted. The good guys knew I called last night, but the bad guys could also have known." JJ thought back, he went halfway to Warner

Springs to use a phone, on a two lane road at night. He would've noticed being followed. In fact he looked up the road during each phone call. He wrote to have Karen ask, "Who called for a limo, if they say Jeremy James Jordan; it has to be the FBI."

She opened the front door again and asked the driver, "My husband's in the shower, who ordered the limo?"

Feigning a look at a blank work order, "A Jeremy James Jordan. Last minute, around nine last night. He called through a mutual friend." Owen could not afford to let on; he hoped the full name would be enough.

"We're almost ready; you can help with the bags." She unlocked the door, and let him in. "With boys you always run late, I still have a couple of things to throw together. JJ came down the stairs with his suitcase in hand.

He sized up the driver and asked about how long to get to the airport this time of day? JJ knew that answer, he had done it for years.

"Wednesday morning, leaving about eight, we'll there by 10:30, maybe 10:15. What time is your flight?" Asked the driver.

"11:45."

"No problem, you'll definitely make the flight. I'll take these to the car."

"I'll carry this one," And he walked out to the car. Once outside JJ quietly asked, "Who are you?"

Owen said, "Agent Owen Tarthman, NSA. The FBI and I have been onsite all night with your other friends."

"Can I call you Owen?" He outstretched his hand.

"Absolutely, can I call you JJ?"

"Absolutely."

The loading went quickly, Owen reminded them, "Lock everything up tight, too many times I bring clients back into a BE crime scene."

"What's a BE crime scene?" Asked Suzanne.

"Why ma'am, that's breaking and entering, people know you're going out of town cause of the limo, they knock on the door after we leave, figure no one's home, and commit a crime. I have seen it too many times. I tell my clients to mind all the locks and dead bolt everything. Just a courtesy." Owen said, exaggerating a drawl.

When the house was locked, the limo departed Timothy logged the report. He delayed the transmission in case something happened while the limo was there. At 8:12 he texted his report to the Order:

> *Subjects all departed in a limo, plate 9U84454, black for 11:45 flight to San Francisco. Caucasian driver. Recorded nothing unusual. End of Report.*

Owen received a call from the FBI agents, repeated the all clear transmission. He called into the back and asked JJ to come forward so they could talk.

67

Lester woke up wondering if he had done the right thing. He had given up the information. It was no more than he had mentioned so many years ago when he was told he was worrying about nothing. He wondered if he had helped or hurt, but he felt it was the right thing to do given the circumstances. If the Nazi's were hanging around Karen's home, she was truly in danger and only the government could help the family. He checked email and was disappointed to see Karen had not answered his email.

He showered, dressed, went to breakfast, and hoped he would see Karen when she arrived. He was invited to sit with some colleagues, it seemed so mundane to talk shop, but he found it normal after a while.

"So Lester, why do you suppose they invited that Hindu to speak?" Asked Dr. Tom Folsom, a practitioner from Texas.

"To present all sides of research I suppose. I understand the meditative process is another form of therapy, Roger probably asked him to share his experiences."

"But certainly there can't be much to it, I mean, has he taken a complicated dementia and cured anyone? It's quackery."

"I don't know if he has, but I don't know that he hasn't. Do you know?" Lester challenged the new speaker.

"His abstract reads like a chapter from a Zen pamphlet. Are you saying there may be some truth to his practice?" Challenging Lester right back.

"Until I hear him speak, I can't say I know. But in my experience, there are many unconventional methods of treatment. If one person is helped, and it does no harm than give false hope, how is that quackery? The patients that do not have better results will end up with you for help. Do you have 100 % recovery in your practice?"

"You know I don't, no one does. It is the nature of the problems; they may be too deeply afflicted to be helped."

"So if I had sat down this morning with the swami, do you believe he would be so critical of your conventional methods, or as I know from my dealings with various healing professionals, he would also say that no method is 100 %, I have helped some."

"Lester; that is the most ridiculous thing I have heard you say yet. Give me Prozac and Ritalin and I can run rings around your mystic healer. Let's talk percentage of success."

"But does a success count if you addict them to drugs for the rest of their lives? What if there are better methods. What if we can quiet the aberrant brain functions with new methods? Maybe the swami has something we can all use."

The dissenter gave up, his logic was exhausted, and he believed it was so self-evident. "Well I can only agree to disagree on this point. I don't want to be closed minded, but this is going beyond my ability to accept as a useful therapy."

Lester had many times upheld the misunderstood, the new treatment, alternative therapies. He knew from experience there were no catch all therapies. He hoped Karen or someone could develop a method to control all functions, complex as it is to do. That would be the closest to a catch all as possible. If a tool could target atrophied matter or recreate stimulation in lesser used matter, many of the dementia could be tested for and therapies generated. But the tools had to exist before these efforts could be made. It was the grail of neuropsychology. But not this year, knowing what he knew.

He finished his breakfast and attended the opening ceremony of the forum; a keynote address by the chair of the Berkeley Psychology Department. Berkeley hosted this year's forum and therefore the chair was given the opportunity to make the opening speech. He spoke of this year's graduate research and followed with a summary of the different speakers and their contributions over their careers. He spoke elegantly of the efforts made in the fields of Cognition, Brain, and Behavior Studies underway at California in addition to Basic Clinical Studies to produce the next generation of practitioners and researchers. Finally, he thanked the organizer and champion of detail, Dr. Roger Hadlyn for putting together what appeared to be the best collection of knowledge of the human mind ever assembled.

Hank and Megan applauded as did the entire room. "He made 20 minutes go by and I wished he never would have stopped." Said Megan to Hank. "I almost wish I had become a psychologist just to be amongst this crowd."

"A good speaker does that to me too. It's interesting to have a knowledgeable person explain complex things in laymen's terms. That is the best of all worlds."

"Honey, Shaolin Koan is going to be in the Empire Room in 20 minutes. Let's find some tea and get in the front row. I want to be as close as possible."

As they rose to leave, the crowd behind them was also rising causing a delay, made slower by the conversations among the participants. Hank was standing next to Lester by coincidence. Hank said, "Hi."

"Hello. I don't know you, are you attending your first forum?"

"Yes, Megan and I are actually here to hear one speaker, Master Koan. He is a tremendous speaker."

"I don't know him personally, but I have read some of his work. Do you practice Hindu or Yoga or both?"

"Yoga. Though I'm familiar with the faith, I choose a path of finding self using the guiding principles. Some would say that makes me a Hindu, however if pressed I would say no."

"So the organized practice doesn't interest you. What is your interest then?"

"Besides having seen the master at the ashram, I have questions I would like to address regarding lights I see in meditation. I'm hoping my way will be further enlightened by master Koan."

"Lights, can you describe them? I should introduce myself. Dr. Lester Warwick."

"I'm sorry as well. I'm Hank Burns and this is Megan Stonecipher."

"It is a pleasure to meet you. You mentioned lights?"

"Lately, in deep meditation, I see light streams that seem to emanate from the body of persons I do not know. I cannot focus on the person or I lose the vision. I see general light all around that seems to come together into a brighter mass of light entering what looks to be the head of someone."

Lester was dumbstruck by the last sentence. "Do you mind if I accompany you to see Master Koan and be with you as he answers?"

"No, not at all. Are you a yogi as well?"

"No, I'm a psychologist. I never had the bone structure to fold." He chuckled.

As they shuffled through the tables and out into the hall, Lester was curious what this young man could see.

68

Owen explained to JJ that he had spoken to Lester Warwick and now had an idea of what was going on but wanted to talk to Karen when the kids were safely aboard their flight. Since they were minors, Suzanne had to go with them but was going to catch up with them tonight in Berkeley.

On the way, Karen stopped to explain to the boys that they were going to see Grandma and she and daddy were going to go to work.

"As soon as the work is done, two days, Daddy and I are going to come to Grandma's and we'll all be together."

"But I thought we were going to Than Franthithco. I don't want to go to Grandmath."

Jeremy was quiet; he had suspicions about the way his parents were acting for the last two days. "Something's going on, isn't it Mom. You and Suzanne and Dad have been writing a lot, like you were trying to keep something from us."

"No dear, there have been some things going on that Daddy and I have to take care of, nothing you need to worry about. It just means that you have to go to Normal, and we'll catch up with you there. It's nothing to do with you silly!"

Suzanne said, "It's going to be a party at Grandma's. Besides you'll be able to play. In San Francisco, you'd be with a baby sitter while Mom worked. Yecchh! Stephan, when was the last time you went to Normal?"

"I don't remember."

"It was two years ago. Grandma wants you to start coming every year. You don't want to disappoint her?"

"I want to go with Mommy!"

"Oh sweetheart, it's hard, but think about the fun you'll have with Grandma when I have to work. It will be two days of fun. Grandpa says he has a hiking trail that has lots of animals to see."

Jeremy said, "Quit whining, Mom says we got to go."

"I'm not whining..."

"Are too..."

"Am not..."

JJ came back after hearing the last of the conversation going on in the back. "Boys, I need you both to go to Grandma's and Mommy and I will see you in a couple of days. I know it's a change, but sometimes in life things change and you can't help it. We're at the airport, so be big boys and let's get you off before you miss the plane."

They parked the limo and situated the kids onto a flight to Chicago where not only Karen's mom would be waiting, but also the FBI. They would be in good hands.

Karen kissed them goodbye and watched Suzanne and the kids disappear down the jet way.

At the last bend, Stephan turned around and flashed a smile to Karen. With a quick wave, he disappeared and Karen knew it was going to be OK.

Owen, JJ, and Karen sat in the gate area until the plane took off, talking about the next steps and discussing what Karen was doing. "What is 1945 Experiment?" Asked Owen.

"I have no idea. Where did you hear that?" Karen asked. It hit her that Lester had asked her to remember 1945, how did this agent know about that unless he had in fact been listening to Lester's conversations or Lester had told him himself. Lester was giving her a signal.

"Dr. Warwick made reference to that name, but it doesn't matter. You are definitely doing something these folks are interested in. Can you explain it to me?"

"I'm trying to design a modality, a tool if you will, to be able to stabilize aberrant brain electrical activity. Imagine a kind of pacemaker for the brain. For the last five years or more, we have been trying to establish what energy levels and frequencies would be most effective. About a month ago we found a very high frequency energy source appear to be enervating fetal brain cells. It looks like the energy is necessary to spark the brain cells to life. We found it when the constant energy lowered when I was over the apparatus. We were able to locate where in my body and it was over the womb, near my baby's brain. We are theorizing that maturing brain cells must have some mechanism by which the energy accumulates in the cells. When the energy levels dropped, it is a loss of a pass-through particle being measured on the net. The only explanation is that the particle is absorbed by the fetal cell. We spent weeks verifying the equipment until we established my embryo was accumulating the energy."

"So much for the holy spirit! So you were looking for something and stumbled onto this effect." Owen thought for a minute. "I have to say I'm absolutely at a loss of what to do with the technology, but I'm sure we can use the opportunity to catch some bad guys. Why was the German surveillance team ordered to wait until eight tonight before they were to act? Is there something happening today?"

"A German doctor emailed me several weeks ago. I answered him and offered to meet him this afternoon at the hotel. At first I thought he was just an interested researcher, but then Lester told me the story." Said Karen.

"You spoke with Lester?!? When? What did he tell you?"

"I met him in El Cajon, he said he was there when the allies entered a concentration camp and he read some documentation of experiments that killed many babies. The papers weren't complete, it appeared to him the Germans tried lots of things and found something that worked on a larger population. He also said he didn't trust anyone with the technology; that it would lead to the world killing itself; so he took some of the key documents hid them and now has people like you following him."

"Owen, I should mention I made a phone call to some Interpol friends of mine, but I don't think they will be helpful, he said they only come in when an agency asks for help." Said JJ.

"Very resourceful, but he is right. Karen, what was your plan when you met with the German doctor?"

"I hadn't thought about the specifics, but I was going to size him up against what Lester painted. I wanted to protect the intellectual property, but now that seems like such a small part of all this."

"Look, let's discuss some possibilities. On the Interpol thing, they are charged only to act officially when international agencies ask for help if borders are crossed when a crime is committed. I believe we have conspiracy here if we can collect the evidence lawfully, and get their help. But for the moment, we need you to get to Berkeley and you need to keep an appearance there is nothing amiss. Do you think you can do that?"

Nods from both of them. "Good, then let's get going and we can discuss the strategy on the way. I have a plane secured; you're flying on Uncle Sam today."

69

The Lear jet touched down at 12:32 p.m. at Oakland's North Field, runway 27 right. They taxied to a fixed base operator where the plane would be serviced and a car was waiting for them. Sam was in the driver's seat, he saw them coming and stepped out to introduce himself.

"Sam Valides, pleasure to meet you." He extended a hand.

JJ took it. "Pleasure as well. JJ and Karen Jordan."

"We're 30 minutes from the Claremont, sit back, relax...," Sam opened the door, went back to the driver's door, and got in. Owen followed his cue and got in the front passenger seat. "I've a call scheduled for 1:15 p.m. with the AD. He wanted you on it." He said to Owen.

There was not much conversation as they whisked north on Interstate 580. Sam elected to take Ashby Avenue and followed it all the way to Claremont Avenue. When they arrived, Karen said to Owen, "We have a room reserved we aren't going to use. It was going to be for the boys. You're welcome to it, Owen."

"Thanks, I'll take you up on it. Get settled in and we'll meet you and JJ in your room at 2 p.m. If the German approaches you before you get there, do as we planned, we need to prepare."

Owen turned to Sam and said, "Let's get on the phone; there are a couple of things we need to order."

70

What do you mean you're with the Jordan's? Have you lost your mind!" the AD yelled into the phone, "You made contact with Warwick! Do know how many regulations you've violated. I'm taking you off the subject and we'll have to talk about keeping your job!"

"Boss, there is something to Warwick's story. The fact of some kind of force to kill embryo brain cells is confirmed by the Jordan's. We have a bigger problem if we lose control of the technology to some radical groups like the *Order*, the *National Socialist German Workers Party*, or the *Blood & Honour*. I made a field call and I stand by the decision. Give me two days to wrap this up and then you make your call." Owen asked.

"Owen, you know the rules here. Never, NEVER contact the subject. There are no exceptions."

"Not even to obtain information of a crime? In order to know what the next move was, I had to know what we were dealing with. If you ask me, Warwick has taken a lot of abuse for no good reason. He shouldn't be a subject; we should support him getting a medal. Boss, give me two days, I will have this wrapped up, and the Director will get two whopping commendations, one for taking down a bunch of Aryan fanatics and another for preventing a catastrophe from happening; the neo Nazis committing murder on a large scale."

"Owen, if you're wrong about this, you'll be wiping up lavatories in the New York Subway, God help me. Do your plans include violating any more regulations? If they are, you're on your own. What's your plan?"

Owen and Sam detailed the actions for the next two days.

It included the FBI and local authorities. The end result was to get the *Order* and if possible, the *Blood and Honour* group to boot; wrapped up in a firm conspiracy case, and protect the Jordan's until the test equipment was secured. For now, the equipment would act as bait. When the *Order* went into the house, the authorities would act; quickly. The good fortune of the surveillance find was going to cement the coffins of Roger Deichmann and Hans Schick. But they needed a way to tie them together.

"I placed a call to Interpol; they are following up on the Jordan call. They returned the call and we have Intel on this Dr. Schick. A loner, very arrogant, grandson of a concentration camp doctor reported to have been involved in human experimentation. No, direct evidence was recovered to prove the grandfather's guilt; he was acquitted in Nuremburg and lived in Bremen until his death in 1977. Some of the papers our Dr. Warwick took were purported to have contained evidence of Schick's involvement and it was OSS and later the CIA who authorized his observation. The fear was that Warwick would continue the experiments." Sam reported, reading off his notepad.

"My gut tells me Warwick is being truthful, he did not want this to come to light. His fear of our government may be over the top, but on the other hand, we took the German scientists and made an atom bomb." Said Owen.

"So do you have all the resources you need?" Asked the AD.

"We could use three more teams from the FBI, they need top secret clearance, and we don't want this to get out."

"I'll make the call. Owen, if it turns out to be a lark, this is going to be difficult to justify upstairs; I hope you're right.'

"I am boss."

"Ok, let's do this. Sam, have you set up listening posts around Deichmann's yet?"

"It's up and we're listening. Still sorting out the chatter, waiting for the hourly update. They should have it at the 2 p.m. transmission. Once we have Deichmann callouts, we will have him cold. There isn't any suspicion as yet; this should go as planned."

"And the surveillance team in Julian? Owen, they weren't cleared for this, what are their bosses being told?"

"I'm recommending they join NSA, very good operators. Maybe we can upgrade them now and that will keep the information in-house."

"That is going to rile up the Bureau, but I will get the decision made higher up."

"OK, I think we need to get ready. Boss, this is really big, this could be the biggest thing since the atom bomb."

"You have two days. Good luck, and Sam, keep him out of trouble!"

71

At 1:55 p.m., Owen checked into the room originally reserved for the Jordan's kids. He pushed open the door and knocked on the adjoining door. JJ opened it. Sam followed Owen in and swept both rooms for transmitting devices. There were none to be found. Quickly and silently, Owen and Sam checked for passive devices. Both Karen and JJ were fascinated by the efficiency and apparent knowledge they displayed.

"This is just like the movies." Whispered Karen to JJ. He nodded.

Sam said "All clear."

"OK, we have a plan; it's going to require a real coordinated effort. You're both still up to it?"

JJ and Karen nodded. "Is it dangerous?" Karen asked.

"I don't think so. The people you are dealing with are professional doctors, the real bad guys are here in Oakland, but not expected here at the hotel. But things are always changing in this business. I can't promise anything, so don't sue me if something happens, but we are here and with enough support to stop almost anything."

"That's not reassuring. What should we do if someone pulls a gun on us?" Asked JJ.

"We don't expect it is all I can say. I can't give you a weapon if that's what you're asking. Look, we'll have agents in every room you're in, stay together and keep your eyes open. If you see something dangerous, know we are watching your every move and we can react. If it gets dangerous, we'll take you to a safe place. It all depends on how Schick reacts to your part of this effort."

Sam's phone vibrated, he walked back into Owen's room to talk to Steve and Bruce in the FBI surveillance Expedition. The *Order* surveillance team texted their 2 p.m. report to Deichmann. They had a fix on Deichmann's phone and they could follow him with Cell Signal Transfer Point if he moved around. It would take a while to pinpoint him, but they could put the phone in a square two mile radius. He is still near his office on E 14th Street and 19th Avenue."

"What was the message?"

Observed nothing out of the ordinary. End of report.

"At 9 a.m., the report contained a request to ID the limousine, have they received any information on the car?"

"They received a transmission verifying the limo was legit. No suspicion. That was at 1:11 p.m."

"Still no other change of plans for the break in, right?"

"No change. If there is anything out of the plan, I'll contact you."

Sam hung up and returned to the second room. Owen completed the placement of recording device on both JJ and Karen. They didn't want to risk using a transmitting device; they would have to make use of long range microphones when Karen was in public places and rely on the small recorders for any missed conversations.

"OK, tested and operational. Karen, when you see Schick, get him to engage in as much conversation as you can. Try to not sit in the corner of any room; we will have other devices recording your conversations in the open areas. If he insists on privacy, don't make a scene about it, but always try to sit first so out of politeness he will sit. We know he is tied to a neo-Nazi group here in North America; we can assume he is sniffing out the details of your work. Do not let on any knowledge beyond the modality."

"I got it. Get him to admit as much as I can. Find out his motives and stay in the open."

"Right, think of it this way, always sit somewhere you can stand and run into safety, a crowd, a very public place. Things like that. Don't run into the bathroom or enter any room unless you know there is another exit too. Rules of the business."

"What do you want me to do?" asked JJ.

"You're the supporting husband. Concentrate on whether he is telling the truth. You can question his motives out of a protective feeling for your wife, but don't go fanatical about it. Just size him up. If he is really just a collaborator in the field looking for a medical device, he will act one way. If he is looking for something more, he will tip it off some way or another."

72

Hank, Megan, and Lester were in the second row, looking and listening to Master Koan. He was slated to speak from 2 p.m. to 3 p.m. The room was almost empty, not many conventional doctors gave much credence to the use of meditative therapies.

In truth, most patients have little patience with the time it takes to have any effect from meditation, but for those who do follow the discipline, it is rewarding. The fact there were few doctors didn't bother Hank; he was going to ask the master questions as soon as his presentation was over. Lester was somewhat disappointed though.

The swami did not sit as he spoke, he did not wear flowing robes or have incense burning. He was dressed in a business suit with his hair in a pony tail. He had a lapel microphone and his hands were calm, fingers touching in front of his belt. He looked very relaxed and patiently waited for the stragglers to finish taking their seats. In a room that would hold 300, there were only 30 or so.

"I am honored to be here and pleased that each of you has taken time to listen to my humble presentation. It is always an honor to be asked to present a point of view for review and to help someone eliminate their pain. Thank you for coming."

"I do not have a medical practice; I am the master of an ashram in Dallas, Texas. I studied under several masters in Hong Kong and in Thailand. My studies in the Tao were forced by my parents, later I found the inner peace a refreshing existence. As I moved in concert with my Tao, others noticed a new peacefulness in my bearing, my demeanor, and the peace I try to bring to everyone in my environment. It was suggested I continue my journey by completing my doctorate in religious studies here at Berkeley and ultimately I founded an ashram in Dallas, Texas."

"The use of meditation is as old as Hinduism, millennia have passed with little fundamental change in the practice and value of it, however in our world; the pace of life has changed. Instant communication, the desire for instant gratification, fast food; for examples, have created an expectation of instant health when one is unwell. The expectation of taking a pill and getting better prevents many from ever getting close to their Tao, to their inner self. I wish to comment on the techniques that exist that, while not being a pill, can help to provide enough glimpses of one's Tao to continue a treatment of meditation that can fundamentally change the overall health of a patient."

Hank and Megan agreed with the comments, it was so true. Lester was interested in this line of logic. He drifted off into his own thoughts of patients who after nominal efforts had such high expectations and were more depressed when no obvious changes resulted. When the pain of emotional disease could not be alleviated by their own minimal efforts, they often turned to a pill.

The master described his program, the starting of habitual meditation. He explained a good habit was as difficult to change as a bad habit. The secret to the system was to change the habitual behavior. This was not really new in the therapeutic world, only how he inventively led his patients to follow the meditative structures he laid out. He spoke for 40 minutes and there were a handful of questions, mostly questioning how he measured the success of his program.

"Has the AMA done any studies that endorse anything you have described here? How are you assuring your patients that the treatment is helpful?" Asked a rather caustic man. Lester turned to him and found Dr. Folsom, from breakfast. Too bad thought Lester; that is one closed mind.

"Good question Doctor. None to my knowledge are completed, but there has been much activity lately; The Mind and Life Institute and the NIH have begun studies. However many professional sports teams have endorsed and practiced the method. In addition, many members of the Texas Legislature endorse the program after their personal practice of the method. There are over 2,000 members of the ashram, those who continue to come after five years of practice. Certainly there are many who have come and gone, they have found other paths to their Tao. I am humbled that so many have found peace and quiescence, perhaps they will find it with other methods, it is for them to say."

"But you did not answer the question. How can you offer a solution in a world where I'm regulated, pay malpractice insurance? I worry that everything I own could be taken with one bad judgment! How can you profess to help the ill with no FDA regulation or costs that place the same burden on you that we all have?"

"I fear you do not understand our program. We do not offer dangerous pills, no surgery; we only offer a new way of thinking and living life. When those who are lost seek help to be found, they have risked nothing and we make no promises. Their way is made better when they find inner peace. We believe most problems are within the spirit and therefore can be cured only from within. If our way does not work, we have promised nothing that we can be sued for. In your world, you try to define the problems and provide solutions from outside, the contraindications are dangerous. You provide medications that possibly cause other unintended side effects. It is the way you have chosen. I have chosen a different way. With your choice, you have your FDA and your insurance costs."

"Are there any other questions?" The master asked. The murmur in the room went up a notch and one doctor stood and left the room. Dr. Foster shook his head and said quackery under his breath and four more doctors left the room.

"I have a question Master." Hank spoke as though it was he and Koan in the room. He had the feeling of a tunnel forming around their eye contact, excluding all other distractions. "In my meditations, I have seen very strong lights, I associate with new life. I find the light is bright one day and not there at all the next. I have an intense curiosity of what the light is. Have you experienced this?"

Without hesitation and without turning away from Hank, the master spoke "You see the life force. It is a rare thing to see it; you have found much of your inner self to have such sight. It is a reflection of your inner peace. It is the pinnacle of meditation to observe the essence of life. Please stay after the presentation and I should be honored to speak to you afterwards."

Quickly, the tunnel of focus disappeared as the master scanned the audience for more questions. "If there are no other questions, I shall be honored if you would take the information provided in the registration packages and in the back of the room. Thank you." He took off the microphone and the room emptied. The swami came over to Lester, "You are a thoughtful person. What do you seek here?"

"I seek to know if others can see the life energy."

73

Karen and JJ went to the lobby, picked up the house phone and rang Hans Schick's room. After two rings Hans was on the phone, "Ja, hello?"

"Dr. Schick, this is Dr Karen Jordan. I'm in the lobby if you have a minute to meet. I have a later commitment, so now would be an ideal time if you are free."

"Dr. Jordan. I will be right down." Hans hung up the hotel phone, sending a text to Deichmann:

She is here, I will meet with her now. I will have an answer for you soon. I still suggest we take the equipment now, while the house is unguarded.

He received a return message:

Talk to the woman, we will move when we determine it is not available for purchase. Herr Schick, I will not remind you again that the plan is made and you are to follow it.

Once again Han's felt the futility of inaction by of the American coward. He deleted the text message with an angry touch of the button and threw the phone into his pillow. He went to the bathroom, groomed himself, tied his tie, and lint brushed his coat. After his meticulous care, he stared at the mirror for imperfection, found none, and left the room.

JJ and Karen were sitting in the hotel lobby, on a couch when the elevator doors opened and Schick came into the lobby. He looked around and saw Karen and walked over to her and asked "Dr. Jordan?"

"Yes, Dr. Schick?"

"Yes. Can we find a place to speak?" He looked around as Karen and JJ both stood. At first Hans did not see JJ, when he turned around, JJ was arm in arm with Karen, and Hans realized he had overlooked the man.

"Hi, I'm Karen's husband, Jeremy."

"Dr. Hans Schick." He did not offer to shake his hand. "Dr. Jordan and I have business to discuss. Could I trouble you to allow her and me to speak in private?" He said directly and coldly.

"Dr. Schick, my husband is quite familiar with my work and I wish him to be included in our conversation." She was as curt and began to lead the way to a table. She was sizing up a self-absorbed personality, one who enjoys being in command.

Han's felt uneasy, this woman dictating terms of their conversation. Unmoving, Han's said, "I insist we have this discussion alone. It is a matter of urgency."

"Perhaps we can converse right here then. What is your interest in my work?" Karen challenged Han's.

"This is unacceptable, we must speak in private."

"Doctor, you may have all the privacy you require. Come on honey, let's grab some lunch." Karen said knowing the predictable result of her blowing off someone who is self-important.

"Dr. Jordan, forgive me, I have not adjusted to the time change. Please let us converse. I have a tremendous opportunity for you; and your husband."

Karen and JJ led the way through the lobby to a table near the door of the bar. There were few people at the bar, the conferences were going on. "Drink doctor?" Said Karen, attempting to command the situation.

"No thank you. Doctor I..."

JJ interrupted, "I'd sure like one."

Han's was becoming infuriated, JJ secretly smiled. "Yes, I'd love a wine."

"So where are you from doctor?" JJ said.

Han's was now confused. He did not want friends, he wanted to buy her technology, or take it. It mattered not to him and his impatience was showing. "Sir, I am not here for a drink or to make banter." He looked at Karen and said, "I wish to make an offer to sponsor your research."

"I am sponsored." Karen said. She was deducing that he was under stress. I wonder what his stress is. "I have never heard of your research, exactly what do you know of mine?"

"Doctor, I have been a practitioner, such as yourself, and have tried to accomplish what you have in your work. Your abstract said you developed a possible non-invasive waveform injection technique for reducing and possibly eliminating the effects of severe brain dysfunction. This has been the emphasis of my work as well."

"But why would I allow my sponsors to be brushed aside to go with you? What is your intention?"

"I am only a kindred doctor who wishes to see the advancement of this technique as quickly as possible. If I can assist in quickening the pace of development, I believe it is mutually beneficial."

JJ asked, "Who is sponsoring you? It takes thousands of dollars to maintain this kind of development."

"My sponsors are of no concern, we are fully funded however. What are you provided with, Dr. Jordan?" Han's was not interested in JJ's interruptions.

JJ had dealt with many cultures, some were curt by nature, some by language barriers; this guy was simply evasive and rude, he thought. "Hon, there's the waitress...Can we get a Cabernet and a Merlot please, Doctor, for you?"

"Nothing." He replied. He was so close, he was tempted to make a call to steal the equipment; this was getting to his patience. "Doctor Jordan, How much funding do you receive now?"

"Ample I assure you. But I'm still unclear if you are clear on what my research is. What do you know of it besides my abstract?"

"I am familiar with your research papers. The advances you seem to have made in the last six months indicate you may be ready to begin controlling brain energy. If this can be done, then it will be of huge significance for the world."

"Doctor, you are obviously an important person, no doubt you are highly respected in Germany; however, I have nothing to offer you. I'm satisfied with my funding, the development will continue at my pace. I'm flattered you are interested, however I'm not inclined to change my funding source, unless you can be more precise about whom your sponsors are." Karen was really fishing, hopefully not too much to set off alarms in Han's mind.

"Doctor, it is a simple proposition. We can offer more than you are getting now. If it is personal credit you seek, it is yours. I speak for several organizations who are interested in this advancement to further humanity, to reduce pain and suffering."

"But who are they? The Bundesarztakammer, German Medical Association?" Asked JJ.

Han's had not considered that this husband probably was familiar with German organizations. "They are interested, yes."

"But not the sponsor. Who are they?" Karen was pushing Han's.

"Tell me doctor, have you discovered some unexplainable energy?" Han's changed tacks, hoping to change the subject.

"It is not my interest to explain any more of my work until you explain who you are representing." Said Karen.

Han's smiled, "Then Doctor, good day."

Han's returned to his room and texted:

The equipment must be taken now.

74

Karen and JJ were shocked by the abrupt departure of Han's. It was 3:25 p.m. The conversation left Karen believing he was not a researcher. He should've been more specific, and he also had megalomaniacal tendencies. His dress and demeanor indicated he was one who felt comfortable ordering people around.

"That man has a very short fuse." Said JJ.

Karen agreed and said, "I want to call Mom, the kids should be landing about now."

JJ pulled his cell phone out and called. Mom said "Yes, they just arrived; there are two young men who showed me FBI ID, we are fine. Suzanne is returning on the 5:20 flight; that will put her back around eight your time. She had a minute to fill us in, is everything alright?"

"Things are fine. We have some pretty good guys on this end too. Thanks for taking them on such short notice. Want to talk to Karen?"

"Yes dear. JJ, keep her safe."

"Will do, here she is."

"Hi Mom, are the boys right there?"

"Heavens no, they saw your father and raced to the car, screaming about who would get there first. Don't worry about the boys, figure out what you are up against, and get it over with. I don't want to worry for long. Love you."

"Love you to Mom." And the line clicked off.

"OK, the kids are safe. We need to find Owen."

"Right here, as he pulled up a chair. Well, that sure set him off. Sam just got a report that Deichmann has already responded to Han's and has texted to the surveillance team up at your place. We plan to have eight guys surrounding your house, if and when they enter the house. Good work riling him up doctor. JJ you were great also. Now we need to put you two in a safe place for the moment."

"But what about meeting Lester? We had planned to meet with him."

"That was if Schick hadn't stirred up the hornets. Now you are expendable, I need to make sure you're safe. I'll bring Lester shortly. He is in a conference room; I'll bring him when he is out of it."

They all loaded into Sam's car for a drive away from the Claremont.

75

Roger Deichmann never liked Han's Schick, too full of himself. He concluded that Schick was a member of the B & H for personal recognition, not to help return the world to a proper order. But when he received the text message, he ordered four teams to meet up with the surveillance van and get the lab equipment. He planned to meet with Hans and see if he was fit for the assignment to take the equipment to the next level.

After organizing the needs of the Julian teams, Deichmann set out to go the Claremont and meet with Hans. It was also coincident with the last of the week's cell phone use. He never used one for more than seven days. He recycled them, had the IMEI number electronically changed and had technicians re-identify them with new SIMs. He had not lasted this long in his covert position by simple oversights; he tried to use very deliberate tactics.

He tossed the old cell phone to his technician who made the porting of his number to the new phone. It would take about ten minutes for the phone system to get the new one operating. He waited and once the swap was done, he left the office. He got in his car and began the drive to the Claremont.

Around the building were three teams of FBI agents. A call was made to Sam stating Deichmann was on the move. Sam ordered one team to follow him.

"Deichmann is on the move." Sam said to Owen. They were in the front seat; they spoke quietly while JJ and Karen were quiet in the back seat.

"Probably not out to the grocery store for an Enquirer. Conjecture, where might he be going?"

"Could be he is just running an errand, but it's more likely the last message he got from Schick is driving him somewhere. To see Schick? To pick up a lotto ticket? Based on that last text from Deichmann to Schick, my bet is to talk to Schick. Remember Deichmann's earlier text; it reads to me Schick may be bossing around the wrong folks." Sam and Owen reviewed the printout of the text transmission:

> *Talk to the woman, we will move when we determine it is not available for purchase. Herr Schick, I will not remind you again that the plan has been made and you are to follow it.*

"It does sound like our man is bucking the system."

"We have one of the bureau following him we'll know soon anyway." Said Sam as they wheeled into the NSA building with JJ and Karen.

"You be safe here. When we get Lester, we'll bring him here. Karen is there anything else you know about the experiment that you haven't told me?" Asked Owen.

"No, I have told you everything I know. But you haven't told me everything; how did you know to find me originally?"

"I can't say how, but I can say it was something Lester said that triggered all this. He called your research the 1945 Experiment. Does that mean anything to you?"

"No, you asked me that earlier."

"I know; I'm just trying to put all the pieces together. If Dr. Warwick was trying to prevent this technology from being used as a weapon, and the Germans are aware of it, I'm trying to keep a lid on it. Who besides you two and Suzanne know about it?"

"About what I am, er; what I was trying to do or about the dropouts?"

"The dropouts."

"Just us. I haven't even told Lester about them. That's why

I want to talk to him. Maybe there is a way for us to continue without the side effect, if it is even related."

"We need to keep the knowledge to ourselves, if this is as dangerous as Dr. Warwick believes it is; we all need to reflect on the ramifications of its potential release. In the meantime, we have some bad guys to round up. I'll be back; I have some errands to run."

76

Deichmann made a cell call to his peer in the San Diego area and said he had a project that would interest him. "The project will require eight man hours." His counterpart agreed it would be possible depending on the time.

"If we start around nine tonight, we should meet our schedule."

"That can work; I will see the project through personally. The proof will come in the ordinary way?"

"Yes, it's on the way."

By use of coded emails, the *Order* had a sophisticated network of information sending and retrieval. This is true of most subversive groups; they had learned that only one method of communication was fatal. They applied the rule of three. Use three methods of communication for all information and the key to it was the randomness. The first was the phone call, a printing job; he had called a printing house in El Cajon. He had said eight man hours, the key was to use terms consistent with the system. Eight man hours was eight men or four teams.

The second bit of information would be coded into an email, protected with PGP, or Pretty Good Privacy, an encryption program. The third would be sent as a fax, using a pre agreed upon time conversion. The email would be the address and the task; the fax would be the time to start.

Deichmann waited for his cell phone to go online and texted the surveillance team in Julian:

BE is on, 5pm

The FBI surveillance tech, Bruce Johnson, in Julian

received an intercept of some cell traffic, a text message that alerted him:

BE is on, 5pm

They waited for any reply from the van, but there was none. "If BE is Breaking and Entering, then this is saying to go in at five. That's only two and a half hours from now. How soon before the other teams get here?" Asked Steve.

"Their last communications indicated around eight. It looks like the *Order* is stepping up the time, maybe they are on to us." Bruce called Owen to report the latest Intel.

Owen took the call from Bruce and asked Sam, "What is the status of the teams getting to Julian?"

Sam explained the situation into the phone and asked, "How long before the teams can get up to the Jordan house?" To the Bureau Chief in San Diego.

"We planned to have them at 6:00 p.m., but it looks like they are running about an hour late. I will have to get back to you. I will do my best to get them free at 4:00 p.m."

Sam hung up and filled Owen in. Owen repeated the information to Bruce.

"That may be cutting it close, but they will be in there for more than a minute. It will take them a while inside to collect everything. I hope they're late." Said Bruce.

"Keep the lines open, if need be we can get some other resources up there. I'll make a contingency plan. Talk later." Owen hung up. "This may be close." He said to Sam.

77

Lester, Hank, and Megan spoke to swami Koan for 15 minutes, and moved out of the conference room and into the corridor. Lester had an idea from all the conversation, but kept it to himself. When Hank thanked the swami for his time and they parted, Megan was filled with pride that Hank had been so recognized by the master.

"Hank, this was so cool that he recognized your abilities. I'm so happy for you."

"The meaning; that is what interests me. The light is the life force he said."

Lester was deep in thought. He was trying to correlate such different facts, he felt an answer to an issue was close at hand, but couldn't quite put it together. He wouldn't have time either. As they entered the main lobby, two men approached Lester and asked him to accompany them to see Dr. Jordan.

"Right now?" Asked Lester.

"Yes doctor." Said one of the agents.

"Hank, I would be pleased to be able to pursue our conversations, please take my card. If you wouldn't mind, please email me so we can talk. I must go at the moment. Megan, it was a pleasure to meet you both."

"Our pleasure, Doctor. I look forward to seeing you soon." Said Hank.

"Where is Doctor Jordan?" Asked Lester of the two men as they left the hotel. "She should be here in the hotel."

"They have been taken to an office where we have orders to take you." Said the agent taking Lester's elbow in tow. It was also the sum total of the conversation throughout the entire drive.

78

Owen drove to meet up with the FBI agents following Deichmann. He was constantly being updated by the Bureau recon team in Julian, by Sam, and was now listening to an ongoing recorded conversation between Lester and Karen going on at the NSA office building. He called the FBI agent following Deichmann and got into traffic behind him. He called his AD and authorized the release of the transcript to him as they spoke of the upcoming multi-force operation. The simple plan was to wait for the Julian bad guys to enter the house and arrest them. Then to simultaneously raid Deichmann's business in Oakland, find and arrest Deichmann and Schick for conspiracy.

Something was nagging Owen. He called Steve Hardin in the Bureau Expedition in Julian and asked him how the *Order* would organize efforts such as this.

Steve said, "Their SOP is covert communication, misdirection, and they rely heavily on multiple means of communications. For instance, Bruce and I have been trying to interpret the latest text from Deichmann. *BE is on, 5pm*. On the face it appears that their operation would be in roughly two hours. As we know, we are an hour from anywhere, let alone the time to get gear rounded up. Unless their resources are fully operational 24/7, I don't believe anyone could move that quick or the guy in the van is going to do it himself with no backup. Since we have nothing but a gut feeling, I didn't call you."

"Steve, I'm authorizing you to report hints, omens, signs, hallucinations, and instincts. Maybe there is a pot of hot water to go through at the FBI to pass that kind of information on, but at the NSA, we have to move smart, quick and with limited information but highly intelligent guesses. I like to keep all options in the discussion, keep on talking."

"Well Owen, we don't know if it is just the one guy in the van who is going to break-in, their SOP says when a replacement comes, they text for the swap of personnel in the van. Their 3 p.m. report was just a report in, nothing new. I don't think anyone would go in alone, that is too risky; so all along we have been assuming there is a more of them on the way. If there are others coming, it will take more than an hour to get here from down the hill. If they are good, they may be organized in an hour."

"Knowing what you know or can find out, when do you think they will enter? How can we find out?

Bruce said "Sir, give me access to recorded cellular communications at the locations I need, I believe we can dig through with some insight."

"I can't do that, there are still laws, but I can put you in contact with our analyst, between the two of you I think we can pretty much find anything."

"There is something else; the text message did not mention any details. Since it only said a time, they have either a play book in the van and the plan is written, or the plan must come in from the outside. If the plan is in the van, then we may not have all the intel that we think we have."

"Keep thinking; I could use minds like yours. I have confirmation from my AD that your teams will be in the saddle in 45 minutes best case, but more likely two hours. Guys, I want to let you both know I think you guys are good; there may be a job change in the near future for you. Gotta run."

79

When Lester arrived at the NSA office building, Lester and Karen hugged each other. Lester asked her if she was OK.

"Fine, just confused. Lester, you seem to know a lot about what is going on, but I still don't believe it is happening, it is too unreal. You were telling the truth, you're not senile. Why haven't you told anyone else after all these years?"

"So many questions, hold on. There are lots of things we both need to know, let's save most of them until this is over. For right now I need you to answer this; do you believe you have found the life force, the energy that enervates newly created synapses?"

"Lester, the evidence is still to scanty, I would not say it is for sure, but it is still not ruled out."

"Do you believe the energy, if it is a life force, could be prevented from going into an area using manmade means?"

"That is more in Suzanne's expertise, but yes, we have the technology to manipulate particles. It is a matter of magnetically steering them around a place, so they don't go where they would have naturally."

"Then that answers your first two questions. Where is Suzanne?"

"She escorted the kids to my Dad and Mom's in Chicago. She should be back tonight. How did you know this could be done?"

"I know that it was done in Bergen-Belsen during the war and I know they were blocking frequencies. They had not identified which specific particles were being blocked. I also know the grandson of one of those butchers has signed up to attend the forum for the first time. We must not let them get to your notes or to the equipment. If they get it, we will find it difficult to prevent the use of this."

"So back in Geneva, at the airport, this was your Armageddon scenario, you believe that if one side gets it, the other side will and there will be much less destruction to the planet, it will be used in a widespread way." Said JJ.

"Yes, Mr. Jordan. But worse, we may find that once it is deployed, any use of it will not just kill, but reduce the brain development to that of children. Thus making the more intelligent group severely advantaged for the final solution, the utter destruction of entire races. The one who emerges will be susceptible to insurgency from within, at the gross extermination of human life. It will pass that the final race will succumb at the suicidal efforts of preventing such tyrannical murderers."

"But that is just one scenario, there could be others; with less dire consequences."

"Jeremy, think back to the history of mankind. Civilizations become great and then decline. Is there something different about ours that will prevent it from suffering the same fate?"

"If we were all of the same race, there would not be competing races, if we could decide to be one people...I see what you mean. It is a difficult task. There will always be diversity. But we have had hydrogen bombs and not used them. We seemed to have lived through that great weapon."

"We did not use it because the winners would not have won anything. They would have a radioactive ball with no natural resources. This weapon is different; it eradicates the people and leaves the resources intact. It has the ultimate deniability. If it is deployed, everyone can blame everyone else. It is a new form of destruction, the power to change the entire world."

Karen interjected "How did you know this back when you were a kid?"

80

Deichmann was taking a very circuitous route, he went south on E 14th Street to Fruitvale, turned east to MacArthur Avenue and taking the onramp onto the 580. He veered to Highway 13 toward Berkeley, took the Chabot College exit, made a U-turn, and reentered Highway 13, exiting on Broadway. He was looking to shake any tails he might have. Between Owen and the FBI team, they could ping pong the chase cars and maintain a visual at all times.

He drove to the Claremont Hotel. Owen passed this ID on to the team's onsite, then called Sam. Schick was still in his room.

Owen wondered why Deichmann would contact Schick. It didn't make any sense to risk blowing cover when Schick couldn't provide any obvious help to stealing the equipment. He remembered Schick seemed to question the tactics, Deichmann had texted for him to follow the plan. If the plan included meeting Schick, why hadn't the intel been able to pick up any conversation. Owen called Bruce and asked if he still had Deichmann's cell signal.

"Hold on, he must have turned it off. The last cell signal was at a site in East Oakland."

Owen was surprised "Deichmann is at the Claremont Hotel. I'm sure he didn't turn it off. He has switched cell phones. Shit." It was 3:25 p.m.; the next report from the Julian van wasn't due until the top of the hour. If he gets a text and answers it, how fast can you get a trace on his new cell?"

It'll take about ten minutes, depending on the number of messages in the clutter."

"So we're blind, we don't think the break-in will be at 5 p.m. and he could be making plans on the new cell for the last half hour. You guys need to be prepared for anything. My guess is they are going to send in a crew. If I were in charge of this op, I would also have limited communications."

81

Lester answered Karen's question. "I didn't, it was written for me to read on the documents. The writers of the documents were all tried and killed. Some of the doctors and practitioners who worked on the experiments were not privy to the big picture; they were given specific tasks only. Schick was a functionary, he heard rumors, but had no idea what he was working on. He passed on the rumors to his grandson. That's why he is here. He will never find what he seeks; he is looking for the haystack and the needle. Poor bastard. Maybe you do not agree with my assessment of the situation, I have spent my life with people like Sam following me. When you announced you had found something, I assessed it to be close to the weapon, and the Aryans began to be interested thereby putting you in danger, I had no choice but to bring them into the picture. Maybe they would not have ever found you, maybe so. I hope our present government will not misuse it now as I fear the government willing to kill hundreds of thousands of innocent people in Japan would have used it back during the war. But I am also old; I'm tired of keeping the company of the NSA. Maybe now I can have privacy."

Karen listened and realized how much Lester had given of himself. He never married, so his wife wouldn't be subjected to his being on the government watch list. He had traveled around the world to become an expert, giving up a family life that would have been his first choice. She empathized with him, and was overwhelmed by the dedication he had shown all these years.

She hugged him and said, "Lester, you could have confided in me, you should not have shouldered this alone."

"No, Karen, I knew I could trust you wouldn't have made a weapon, but I didn't know if you wouldn't have had the treatment other friends of mine have had to endure, although most don't know it. Did they tell you how they got involved? No, I bet they didn't. After many years of coincidental contact, coincidental happenings, I began to experiment. The logical puzzles finally began to show that my whereabouts were known to certain people, glances from similar looking faces were obvious after a time. The NSA went to you because I mentioned the "1945 Experiment" in George Gordon's office in context with your name. Tell her how it works Sam." Lester turned to the agent.

"I'm not authorized to comment to any of this conversation."

"I have heard that many times from the similar faces I've seen. I hope it all turns out well in the end, but I couldn't let you and your family be injured. What are the plans for us now? Has anyone told you anything?"

"Agent Tarthman believes this will be wrapped up tonight, the plan is to arrest them all; here and at home. The plan is for me to finish my presentation tomorrow and then leave here for Illinois to be with the kids."

"Are your children safe in Illinois?"

"I hope so. The FBI and my parents met them at the airport. Thought it best to send them to family until this sorts out. You're welcome to come with us."

"I appreciate the offer, but I have made other plans."

JJ asked, "What are you going to do now that this is out?"

"I don't believe this is over, even though the secret is out, now it will require more work. Someone needs to come up with a means of detection. But I will be in touch; I won't be a stranger any more."

"What about dinner? It's almost the dinner hour, Sam, are we allowed out to dinner?"

"I'm authorized to say we will provide a dinner, escorted, at the restaurant of your choice."

"How un-government-like...," Said JJ...

82

Deichmann received a text message from El Cajon.

The job will be completed on time.

He called his office and asked if the fax to the printer was sent; he needed to know the proof went out as planned. Yes, it had.

Deichmann contacted his superiors and told them of the opportunity that had fallen into his lap. He was ordered to acquire the technology, bring it to a location in Northern California, near Redding for evaluation and to bring the German doctor to help decipher it.

Deichmann voiced his concerns regarding the German doctor's apparent ego and demanding character and was quickly rebuffed for making such remarks about a fellow patriot. They will see for themselves he thought. For now, he needed to find the doctor and plan to get him and the equipment out of Oakland. He didn't want to do it electronically; he could only pass this on personally. It reduced the opportunity of being overheard, and misunderstood.

He texted Hans:

Meet me in the lobby in 10 minutes.

Hans felt the vibration and was surprised he was summoned so quickly. Maybe I am incorrect about this man; he has taken action so quickly. He texted back:

Ja

83

Bruce didn't have a fix on Deichmann's phone, but he did have a fix on Schick's. He traced the data back from Schick's "Ja" to some texted messages that seemed to make sense. He scrolled through the data and found the packets containing "Meet me in the lobby in ten minutes." He traced the packet to an electronic signature of a device and looked up the IMEI and tracked it down.

"Owen, its Bruce, I have Deichmann back. He just texted Schick to meet him in the lobby."

"Great job, I hope we didn't miss something valuable, but my gut says we did."

84

Hans came down finding Deichmann standing outside, in the hotel driveway near the front door. He had his back to the lobby camera Owen noted. Hans came outside and they got into his car and sped away. There had been no opportunity to target the car with any devices. They would have to do this the hard way Owen thought.

"Subject departing, Left out of the hotel." Said Owen to the chase cars.

They began tailing Deichmann. He ordered the FBI agent onsite to stay at the hotel awaiting further orders.

Owen spun his car out of the lot and began his leisurely pace. When Deichmann made a turn, the handoff would be made by the lead car, he would take the rear position, and the second car would pull up two cars behind the subject. It would take someone very clever to figure out this method of being followed. Owen and his team were pros at it.

It appeared to Owen they were going nowhere; he surmised it was just a chance to talk in private with Schick. They went south on 580 to San Leandro, turned around and made their way back to the hotel. The trip took 40 minutes.

"There was no 4 p.m. report from the van. I believe your right about radio silence. Steve did a recon on the van, still only one occupant." Reported Bruce from the FBI Expedition in Julian to Owen.

"Thanks Bruce, keep me up on Deichmann's phone location."

"It looks like you guys just did a loop of the East Bay."

"That's exactly what we did, to come back to the Claremont. Will keep you informed."

"Hey Owen, there is chatter from a phone in El Cajon, it says "the job will be completed on time." We set up a net to sort all chatter from the San Diego area, you never know. This one came, from a printing company in El Cajon. We're following up. Could just be a real print job though."

"OK, keep on it." Owen left his car and walked into the hotel lobby. He hoped to hear anything Deichmann and Hans were talking about. They weren't talking. Hans got out of the car. He appeared red-faced angry. He did all but slam the car door and storm into the hotel lobby. Whatever he had talked to Deichmann about, it must have been disappointing to him. He is just not having a good day Owen thought.

The plan was not to pick up Deichmann until the operators in Julian were picked up. They had to keep up with him until then, but now they had his new cell phone identified, at least they could track some of the intel that way.

Hans went straight to his room. Owen followed him to his floor and feigned forgetting his own key.

Hans hardly noticed Owen; He slammed his door and left Owen in the hall to wait for Hans to come out of his room. After waiting 20 minutes in a service room with the door ajar, he updated Sam and checked in with Bruce.

Bruce called Owen, "Deichmann's texted a 619 phone number, the text reads:"

Advise earliest opportunity of acquisition of package, execute immediately if possible.

"I tracked the IMEI to an address of a printing company in El Cajon, California. Steve is running the business for known affiliations. I have also started monitoring traffic from that phone."

"Things seem to be heating up for our friends. Stay on it." Owen hung up with Bruce and dialed Sam.

"Sam, it looks like the bad guys may get into the house

before we have our men onsite. Ask Karen if she took the research notes when she left. I'll wait..."

Karen took Sam's phone, "What do you want to know?"

"Karen the bad guys are moving faster than we had planned. When you left this morning, did you take all your notes and papers?"

"Yes, I did."

"If they got the equipment out, is there enough data for them to create this weapon?"

"We dumped all the software except for the backup drive and that went to Illinois. If they get just the hardware, it will not be useful to them; in this case the works are mostly in software. It took us years, I suspect it might take them as long. Why, how close are they to the house?"

"I don't know how close, but the head bad guy just gave the go order; and our doctor Schick looked awful mad when he got out of the car."

"Karen, I gotta go. Let me talk to Sam again." He clicked off.

Sam was on the phone for about five minutes, getting an update no doubt and planning their next moves.

85

Sam took them back into the dark room where Lester had spoken to Dr. Fitzsimmons the previous night. Sam said, "You deserve to know some things." He logged into a computer station. On the screen were notes, transcriptions of reports in real time from several analysts, surveillance teams, chase teams, and Owen. "This is a feed of all the intel that is available to Agent Tarthman and authorized case agents. With up to the minute intel, we can know many things all at once."

12:41:12z—From AC619 EL Cajon, CA Printer's phone
Txt—"Product still being typeset, estimated completion 5:45 soonest. Am expediting"
12:42:35z—To AC510 Oakland, CA Deichmann
12:44:28z—Reported by S.CA FBI: SV 1245—auth 1221234
12:45:01z—Ack OT
12:48:23z—From AC510 Oakland, CA Deichmann
Txt—"expect product by midnight, will personally deliver."
12:50:48z—To AC530 Dunsmuir, CA NOID
12:51:34z—From AC530 Dunsmuir, CA NOID
Txt—" roger. will the doctor be with you"
12:52:27z—To AC510 Oakland, CA Deichmann
12:52:53z—From AC510 Oakland, CA Deichmann
Txt—"One way or the other"
12:55:01z—Ack OT

Sam explained, "The time is in Zulu time, that is, Greenwich mean time. It is time stamped and coded to a case. Right now, there are several cell phone text messages we have identified and are intercepting. There are many acronyms; AC is area code; txt is the text of a text message if printed out; NOID is No Identification; SV is me; OT is Owen. By keeping the chronological data stream and a search capability of the record, we can get information when we need it to help solve a crime, prevent the prosecution of one, and create an evidence trail."

"Dr. Warwick, you knew that we were around and had some knowledge of your behavior, but not exactly how. This is the method we employ. We call this Intel stream the "frame". The intel comes in live. You can get fixated on the process if you're not careful; it runs all day long, creating a string of relevant data on subjects, suspects, and ops in progress."

"This is what is available to you guys" JJ was amazed.

"Only with a court order and only for legitimate ops; operations. From here the story is told, recorded, and kept for the record and review. Each input is coded to a particular op. This is just a list of the intel registered to this op. If you were to look at all the data, it looks like a matrix screen saver, streams of data. The trick is getting the data coded to the correct op." He looked back at the screen.

12:59:49z—AC530 Dunsmuir, CA ID Bait and Tackle store, 334 Eureka St. registered to Robert L. Johnstone ID B&T
01:03:33z—Name: Robert L. Johnstone, AKA: John Stone, John Roberts, Rob Johnson. Record: Convicted—Murder, 1967, released 1989, Charged with hate crimes, 1998, acquitted, Charged AB 1999, Pending, Known Affiliations: Aryan Order, Address: 13b Shasta Court, North Dunsmuir, CA
01:06:22z—From AC530 Dunsmuir, CA B&T
Txt—"use west entrance"
01:06:56z To AC510 Oakland, CA Deichmann
01:07:21z—From AC510 Oakland, CA Deichmann
Txt—"roger"
01:07:48z—To AC530 Dunsmuir, CA B&T

As Sam typed his acknowledgement into the computer, he picked up the phone and with his mouse he blocked and printed out the text starting with the first Deichmann contact, wrote a note to include the local Northern California FBI for evidentiary opportunity and possible conspiracy with a subject named Robert Johnstone. "Please acknowledge capability ASAP. Log to the frame."

"Should you be allowing us to see all this? I mean, this is looking like it is illegal or something." Said Karen.

01:08:34z—Ack OT. F/U N.CA Bureau
01:08:54z—Ack SV
01:13:23z—N.CA/FBI: Ack, file 3W33443T2—Johnstone, expect team availability, Ch 56b, Yolo.
01:13:48z—Ack OT

"Not at all, Dr. Jordan. This is all legal, if we didn't do this, the crime rate would skyrocket. Little does the public understand of the dangers out there or the efforts we make to keep us all safe. Think of this as an anti-virus for humans. I don't mind showing you; the danger of showing your capabilities is that your enemy will change their tactics to reduce your effectiveness. Owen authorized me to show you this to let you know you are trusted people."

"But how can it be legal to have hounded Dr. Warwick all these years. This is a violation of his civil rights it seems to me!" Karen burst out.

"Dr. Warwick violated the law, he could have been tried and sent to prison for treason. Because he was also instrumental in the successful prosecution of several of the SS medical corps based on his personal involvement after the concentration camps were liberated, the US government opted to watch him instead of sending him to prison. I wasn't there sixty years ago, Dr. Warwick, maybe you can explain to Dr. Jordan why you broke the law. Regarding using the frame to perform the job of following Dr. Warwick, without this system, you were in grave danger, you might have been killed, the bad guys get the technology and somebody uses a terrible weapon against helpless people; all with no one noticing. I believe in this system and in what we're doing."

"He's right about me and logically correct about the consequences of life without these means, but unless the power is checked from being absolute, it will corrupt absolutely."

"So true. This technology is not exactly publicly available for the reasons I stated before, but there is always oversight. Each case must be authorized before an ASCII character can be printed into the frame. We believe we have the best of the situation, there is oversight, and mostly we get the bad guys. Sometimes we can't get a warrant and the bad guys win, sometimes the intel is not enough and the bad guys win. We are constantly reinventing and discovering tactics to make us better good guys."

86

Owen wondered what Deichmann had said to Schick that infuriated him so much. Since Schick had not communicated to anyone, it must have been personal. They did not seem to be on the same page from the earlier communications.

He surmised if Schick has spent years trying to find this technology, chased it across the world, tried to negotiate getting it from Dr. Jordan, it was likely that Schick was told he wasn't going to get it. Deichmann and the *Order* probably were going to keep it themselves and the good doctor was angry that he wasn't going to get to take it home and play with it. That would explain the intercepts. He toyed with the idea of using Schick's anger somehow to help provide evidence against the *Order*. He wondered if Karen would be willing to help.

Deichmann returned to his office on E 14th Street, with two other agents covering the building, Owen debated his next move. He called Bruce in Julian requesting an update.

"No one else has approached the van and it looks like they have gone to radio silence. There hasn't been any other communication in or out of the van. The back up teams from the San Diego office are enroute, but still 50 minutes from arriving."

"That is cutting it close; the last intercept indicated they would be there at 5:45. Are you guys going to be able to handle yourselves?"

"Hey, I might be a nerd, but I still hold a marksman medal for pistol. We'll do fine," Said Bruce.

"OK, but if it gets too out of control, wait for backup. There is not a lot in the house that will help the bad guys, even if they get it. We know when and where they are taking it, it may be prudent to let them get it and then catch them in Dunsmuir."

"Dunsmuir? Up by Redding?"

Apparently that's where the *Order* has a cell. Intel indicates when they get the equipment, it will be taken there. They planned a midnight meeting to hand off the equipment."

"That means they will have to fly, there's no way to drive there that quickly. I'll call in and have the flight plans reviewed and we'll follow any VFR flights."

"It looks like Deichmann is going to personally deliver the equipment, so we'll stay with him. Check for flight plans to Dunsmuir and to Oakland from the San Diego area. Hopefully we can get others in the *Order* and confiscate some of their resources to boot.

Owen called Sam and discussed the plan to let Deichmann's people succeed in breaking in, taking the equipment, following the flight and having the task force concentrate in Dunsmuir for the arrest. Sam agreed.

"I'll call the AD, coordinate the sting up north. What have we got on Schick? How can we round him up as well?"

"I'm working on that. So far we just have a text message tie to him. But if we can get him to a breaking point, maybe he'll expose himself. Let me talk to Dr. Jordan again."

"Here she is."

"Hello?" Said Karen.

"Dr. Jordan, I need to confirm a few things. The equipment is not the whole experiment, the software is the meat of it, and that is secure; correct?"

"Yes."

"If we let the bad guys complete their break-in, take the equipment to Dunsmuir and arrest them in the act, we can corral a whole lot more of them than just the bad guys in Julian. Are there any other risks you can think of, in case we lose them somehow?"

"It's just that they are breaking into my home. I really don't want them in there. But there is a greater good I suppose. But as far as the experiment, they would have to write a lot of code that Suzanne and I developed. It'd take them years, if they could even get the hardware going."

"Another thing, it looks like Schick and Deichmann are not seeing eye to eye. I'm thinking Schick is going to be odd man out at the end of this. He came trying to get the experiment home for his group, the *Blood and Honour* sect and the *Order* here is trying to make a grab for it. If I'm right, this will make our doctor see red. Can we exploit this to try to uncover his motives and his contacts elsewhere?"

"You mean can his emotional state of mind be used to get him to make some mistakes? Yes. What do you have in mind?"

87

Hans entered his hotel room, threw his jacket onto the bed in tremendous frustration. "The impudence of these cowards!" He thought. "To try to keep the technology to themselves, they do not have the resources, they do not have me." Hans was livid, Deichmann told him the equipment was not going to go to Germany; it was going to go to some small town in northern California. From there the Order would continue the development.

Hans sat and reflected on the years he spent, the time he wasted, only to have the power taken out of his hands. He knew he could notify his associates in Germany, but it wouldn't do any good. There were not any agreements between the *Order* and *National Socialist German Workers Party*, the *B & H* in Germany. He debated whether he should call anyway; perhaps he might convince them to retake the technology from the American cowards.

The more he debated it in his mind, the less he remembered the rules of participation and the oath to secrecy. The more infuriated he became the closer he came to calling the Bremen Chapter leader, pleading his case, forgetting any contact outside of the secrecy of the random meetings was strictly prohibited. He rationalized the urgency of the situation surely must outweigh a call to plan a coded message. At the meetings, he was the representative assigned to find this energy. When they knew he found it and it was being taken away from them, perhaps there were more resources that could be used. He convinced himself to make the call.

"Herr Schroeder, Ja, this is Doctor Schick."

"Why do you call me at such an hour?" Asked the man curtly.

"It is a matter of urgency. We need to communicate on an important matter."

The man gave Hans a number to text and said, "In ten minutes," He hung up.

Hans began typing a text message "I have discovered the technology we are looking for in America. As planned, I contacted friends to aid in obtaining it. They are not aiding us, but keeping it for their own purpose. This is not acceptable. Contact your peers and demand we retain the technology. These men are uneducated cowards."

His phone beeped when a text came through:

"You know the rules and the penalties for such a breach of communication"

Hans blanched, he did know; ex-communication, total denial, overt actions to discredit, and even killing the member who put the organization at risk. He replied with the text he had written:

"I have discovered the technology we are looking for in America. As planned, I contacted friends to aid in obtaining it. They are not aiding us, but keeping it for their own purpose. This is not acceptable. Contact your peers and demand we retain the technology. These men are uneducated cowards."

Certainly this would get his attention and relieve him of the burden of worrying the ramification of this breach.

Several minutes went by; Hans began to worry why Schroeder was not replying. After 15 minutes, Hans was beside himself. He reread his text with a clearer mind; he determined that it was not incriminating. It did not admit anything, but he worried. He and Schroeder had known each other since Hans was a teenager. They had come through the ranks of the Aryan training together. He knew Schroeder would not think he was exaggerating or that this information was incorrect. Why doesn't he answer?

88

The frame came alive with information as Sam looked on. Owen was instantly alerted:

01:53:33z—From AC510 Oakland, CA Schick
Txt—"I have discovered the technology we are looking for in America. As planned, I contacted friends to aid in obtaining it. They are not aiding us, but keeping it for their own purpose. This is not acceptable. Contact your peers and demand we retain the technology. These men are uneducated cowards."
01:53:48z—To Bremen, DE, +4904213080130, NOID

Sam typed into the computer terminal:

01:55:32z—RQT Interpol, ID: +4904213080130, AUTH QW33453321TRQ
01:57:01z—Ack Interpol.
02:02:12z—Reported by Interpol, BREMEN, DE, ID Name: Herbert Schroeder, Record: none, Known Affiliations: neo-Nazi Blood and Honour, Address 24 Finsteinstrasse, Bremen, DE. ID Schroeder
02:03:44z—Ack SV
02:05:21z—Ack OT

Lester looked over Sam's shoulder when Sam seemed to spark up.

Lester said, "I know this name. I believe he is known to be an Aryan.

Sam looked up, "How do you know Doctor?"

"In my travels, as you well know, I have many professional acquaintances, but I have also been researching the family members of the butchers of the Holocaust. I had hoped my simple inquiries would help me to pinpoint when and where the weapon might likely show up. That is how I'm familiar with Dr. Schick. In my research into the Aryan organizations, this man is a member and leader of one in Germany. I have a dossier of him in my computer."

"Where is your computer now?"

"In my room at the Claremont."

02:10:32z—Reported by Interpol, Schroeder arrested on conspiracy charge at 02:00:12z, advise further requests.

Sam was surprised. "Look at that, this guy Schroeder just got arrested, talk about timing."

Owen called Sam almost immediately, "Well things sure move fast at Interpol. Can you request further information from them? Advise them of the status of the investigation here. Perhaps they would be willing to help sting our doctor Schick since they have the phone used to talk to him."

Sam said, "I don't think we can do that boss, entrapment; remember."

"OK, at least find out what Interpol has on him. Didn't the frame say there was no record? It is bad timing to have the connection between the skin heads severed when we could have tied them together."

"I have a request in now."

"Maybe now would be a good time to cause a small stir in Schick's life. I would predict he is expecting a reply. If the reply were from a different quarter, maybe he'll roll on his friends."

"Go for it boss. We are going to get out for a bite."

"Roger. I'll be in touch." Owen said as he hung up.

Sam typed into the frame:

02:17:22z—SV: RQT Interpol: Advise exact charges and affiliation with Dr. Hans Schick of Bremen. Is arrest coincident with present investigation?
02:18:54z—Ack Interpol

JJ remarked, "This is just amazing. So you guys just communicate all the time as though having a conversation. I have a hundred questions, who sees what, how are shifts handled, when you sleep is there someone else covering?"

"Mr. Jordan, I was authorized to show you how we see intel. I was not authorized to explain anything else." Said Sam with some finality. "I'm authorized to take you to dinner."

"But how will you get the answer from Interpol, if it comes while we are eating?"

Sam pulled his sleeve up and showed a watch-like device with a scrolling screen, the frame. "It vibrates with every new input. If I take it off, the DNA sensor on the back of it will turn it off. It requires DNA verification, minimum heat of a normal temperature, a login, and a pulse to operate. If you were to try to wear it and not put the correct code into it, in five minutes, the C4 charge built into it would blow your hand off."

"Oh." Said JJ. He pulled Karen to the opposite side of his body as though to shield her from an impending explosion.

Sam smiled.

89

A knock at Han's door startled him. Owen knocked a second time and added some extra force. Han's looked through the peephole. He did not recognize this man. He was nervous, 22 minutes since his coded text to Schroeder; this could not be a response. It must be just a coincidence. He gathered himself and said through the door, "Ja, what is it?"

Owen knocked on the door again, with another notch of force and faster beats.

Hans couldn't be sure who it was, he didn't answer. He said, "Yes, who is it?"

Owen did not say a word. He continued knocking and Han's became very worried. Who would not answer my request, but I have let him know I am inside. He went to his luggage and took a bottle of cologne, the only thing that he thought he could use as a weapon and opened the door slightly. "Yes, who are you?" He demanded.

Owen held his Government ID in front of him and said, "Owen Tarthman, United States National Security Agency. I would like to ask you a few questions."

NSA, how, why, he began to think of the message and whether the Aryans in Germany had not given him up. He was confused and suddenly he felt defeated.

"Please open the door Dr. Schick; I need to speak to you."

Owen was now recording; this information would be up linked to the frame, for ready transcription and use on the case.

Han's opened the door, walking slowly back to sit on the edge of the bed. He tossed the bottle back into his suitcase.

Owen fully opened the door and looked through the gap for someone behind the door. He walked slowly in and surmised that this was a good time to try to get him to turn on his friends. "Herr Schick, Herr Schroeder has identified you as belonging to a neo-Nazi organization conspiring to steal technology from a foreign country. I have orders to take you into custody on the charge of conspiracy to commit a felony. You have the right to remain silent. Will you come peacefully or will my back up have to help restrain you?"

Glassy eyed, Han's said, "I have spent years in dedication to the ideal of a supreme Aryan world. How could I have been so easily deceived?"

"Herr Schick, understand you have the right to remain silent and anything you say, can and will be used against you. If you help to prevent the continuation of a crime being committed, I am authorized to offer you assistance from the full force of the sentence you would likely receive for participating in a conspiracy. Do you understand sir?"

Hans changed expression when it dawned on him, I can get Deichmann; I can see that coward into the pits of hell. My allegiance is with the true Fatherland, not the cowardly American copy cats. "Ja, I understand. What would you like to know?" Knowing he would limit his participation to what he thought could be used to bury Deichmann.

While Owen was glad to be able to exploit a moment, he was surprised Schick would turn over so quickly. Deichmann must have really insulted this guy.

90

At 7:20 p.m., Sam, Lester and the Jordan's finished dinner and were heading back to the hotel. The intel was that Deichmann was on his way to Dunsmuir Municipal-Mott Airport with a woman in a Bonanza A35. Schick was in custody and initial reports were he was telling the local FBI that Roger Deichmann was chapter leader for the Order. At 6:14, two men showed up at Karen's house, took 20 minutes to carry out several computers, the net and interferometric dishes from the exterior of the house, and loaded them into the carpet van. The FBI agents had arrived in Julian in time to prevent the break in, but were ordered not to interfere with the break in, follow the perpetrators until they dropped the equipment off, maintain surveillance until the word came out that the arrests in Dunsmuir were starting. The sting would be complete and coordinated.

Owen was on his way to meet the Jordan's at the Claremont. Sam told JJ his call to Interpol had sparked the arrest of the guy Schick called. He was under investigation for some covert activities in Sweden and Norway. That made JJ feel somewhat better about time he spent calling. Sam confirmed Schroeder was being investigated and in addition to the Schick call, there were other activities that prompted them to make the arrest. The Schick call further indicted Schroeder and Schick would not like his accommodations in the future.

They took a table in the bar. Owen had been in contact with his boss, the Assistant Director and any breach of conduct was chalked up to the use of good judgment in the field. Owen finished his call to the AD and entered the bar. He sat at the table with the Jordan's, Lester, and Sam, ordered a Stoly with rocks and olives. The conversation was about the frame and its use.

"The technology is not enough to catch the bad guys," Owen explained to the group, "Without the human element. The frame is a tool, but it can't make judgments, it can't use intuition and make any decisions."

"So that is how you justify spying on people?" Karen asked.

"Yes. But the people we spy on are implicated in some crime. I'm surprised you would take that tack Dr. Jordan. What is the alternative? To let the crime happen and try to piece evidence together, after the fact? We have a responsibility to protect and serve. To protect has an obvious subjective twist, but the law limits how we can utilize the capabilities we have."

"Well, I agree crime should be stopped before or while it is ongoing, it makes the stronger case for prosecution. I just shudder to think my phone could be so easily traced and transcribed." Said JJ.

Lester said, "I have issues with its use. My life has been spent, with no real freedom, though it is hard to argue that without it in this case, greater damage would have been done. The greater good, isn't that the common retort of tyrants? Absolute power still corrupts absolutely; we should always be wary."

"In your case Dr. Warwick, I can't begin to empathize with your experiences, neither will I apologize; but I want you to know in our conversation, I recommended to the assistant director that you officially be removed from the watch list. The decision will have to be made at higher levels, but the AD agreed with my recommendation. Furthermore, I have submitted your name and recommendation as a recipient of the Citizen's Medal for your assistance in aiding in the investigation."

"A medal! You should do much more than that!" Karen said.

"I don't want a medal; I only want to see that the weapon doesn't materialize. Since you have discovered it, my fear is it will be found by another. Agent Tarthman, how does the US Government plan to keep this secret?"

"That's a decision way above my pay grade doctor. Since this energy can be detected, Dr. Jordan it occurs to me that another technology could be devised to protect anyone from it. Perhaps you would like to be the person who would help develop it."

"This has all been so foreign to my lifestyle. I'm still as passionate to continue the work for the modality, if in conjunction with that, I would be interested, but only if my work can continue."

"Karen, what are you going to talk about tomorrow in light of the recent events?" JJ wondered.

"I will not commit to anything that would alert anyone else who might want to continue threatening us, but I do want to report progress is still possible for a modality. I think there are two other significant considerations resulting from all of this. What is life if it merely is a biological interaction of the biology and the universe and how we can prevent it from being used as a weapon? It will take a lot of work to continue while walking on the fine lines between religious resistance and the intent of evil people. Lester, what would you recommend?"

"I recall warning you years ago that trying to find God would not lead to a good outcome; little did I know how close you would eventually come. I also want your modality, as much as you. I support your continuing the work and want us to carry on together so there is a responsible check and balance. I'm more educated myself, so there may be other possibilities of monitoring deployment, but I need more research and I want us to work together."

"So I should do the presentation tomorrow and leave a positive spin. Mr. Tarthman, I know this isn't over yet, but before I don't have a chance to say thank you, thanks for protecting my family. Even though it goes against my intuition of freedom and the apparent illusion it really is."

"It is my pleasure to be on the winning side, though it doesn't always work out that way. I do have a question for you, not related to events, connected to the technology. I'm not a religious man, but I guess I have a belief in God, an unseen spirit that instills a soul into the body. Do you think what you have done will scientifically disprove that?"

"I haven't put much thought into it until now, but being the pragmatist I am, I can certainly see the logic of it. Evolution produces changes in all organisms based on the environment it lives in. On evolutionary scales, it is difficult in our short time as intelligent beings to assume we are above the same evolutionary effects. Because we learn, it may be this will become just another lesson in the history of homo sapien; that the cosmic energy is collected by a specific cellular development in our brain synapses that exploits this effect early on and we called it the soul and give credit to God or some higher power. Just because we didn't know does not mean it didn't exist. The obvious social ramifications could be traumatic to those who place such high regard in this specific kind of faith. Are you asking me to help you come to terms with this dichotomy Owen?"

"No, I can always displace my beliefs that God is the one who sends the cosmic radiation, but the icons will likely change at church."

Sam chided in "I'm Catholic. This concerns me a lot. I do go to mass, I believe in the trinity. But if what you are saying is true, what is the trinity but a myth! I can't accept your explanation. It is easy to take shots at faith and those who practice it. I never bring discussions of faith into work, but this is very disturbing to me. There are still unknowns about what you have done, why do you propose what you believe to be the case must exclude God? I don't accept your answer; it is just that you don't have the strength of character to keep faith. That's what I believe."

Lester was in his element and his quick observations led to his final comment, "What an interesting commentary, here are intelligent people, sitting around a table discussing intimate beliefs, in light of clear scientific evidence. Are the facts not plain enough to overcome prejudice and blind beliefs? Agent Valides, you are right to stand for your beliefs, but you are also an observer of evidence. What do the facts lead to given all the evidence you have?"

"That is a disbeliever's argument. The evidence in most cases has multiple meanings. You're all pragmatists and obviously disbelievers of the true faith. I'm not so vocal as to remind you that your souls are at risk, but there it is. It is your choice. I should remind you there are more believers in the world than not. The evidence is already in."

Owen recognized any further discussion would become personal and he knew Sam well enough to try to stem the impending discussion. As he was about to say something, Karen's cell phone rang.

91

Suzanne had spent the return flight wondering what was going on with Karen and the Germans. She had made sure to protect the hard drive, giving it to Karen's father and telling him not to give it to anyone. With her laptop going through calculations, she spent the hours flying back to Oakland trying to determine the feasibility of using the technology on a mass scale.

The plane landed, it dawned on her she would have to catch a cab somewhere and didn't know where "where" was. As she was walking up the hallway, she passed the security line and into the main hallway, a limo driver carrying a sign with her name on it approached her.

"Dr. Vlavich, I have been expecting you. Please follow me."

"Who are you?"

"I was told to pick you up."

A non-direct answer startled Suzanne and she again asked, "Who are you and where do think you are taking me?"

"I was ordered to pick you up and take you to Berkeley. That is all I know."

"Hold on, I need to make a phone call." She said and walked away to get more privacy. She called Karen's cell phone.

The man walked behind her and took the phone out of her hand and then Suzanne knew this could not be a good guy. She turned to look for help when an Oakland policeman was only 10 meters away. She screamed, "Help! This man is robbing me!"

The cop turned and the limo driver dropped the phone

and made a dash to the door, disappeared into the parking lot and vanished from sight.

The cop radioed an ID and approached Suzanne and asked if she was alright.

"What happened?" The officer asked.

Shaken but relieved, Suzanne said, "That man tried to steal my cell phone, I turned and saw you."

"We have two officers going through the lot, if we catch him, do you want to file charges?"

"No, that's alright."

"I need to file a report, what is your name?" The officer asked.

Suzanne went through the report with the cop and tried Karen again. "Karen? Yea, I am at the Oakland airport. Did you send a limo driver for me?" She explained what happened.

"No, I didn't. Are you alright?" Karen asked.

"Yes, just shaken. Where do I catch up with you?"

"Hold on, let me ask." Karen briefed Sam. Sam got on the phone and told Karen to have Suzanne wait for an agent to pick her up. It would be a man named Robert Redford.

"That's kind of dramatic isn't it? Couldn't you guys be more realistic?" Karen said to Sam.

Sam looked at her and said, "That's his real name."

Karen apologized and got back on the phone and instructed Suzanne to wait for a man to pick her up. His name is Robert Redford.

"I hope he has the same looks!" Suzanne said.

The frame reported:

04:25:48z—OAK PD: Suzanne Vlavich, attempted robbery
04:29:33z—OAK PD: Suspect lost

Karen read the wristband on Sam's arm. "That is simply amazing."

"It's all good when it works. But like Owen said, it works best when people are using the information with good judgment and experience. Look, I'm sorry I was getting hot a minute ago, my job is to protect the innocent, and not determine if they believe what I do." He said looking at her. He really didn't like to make any op personal; it took away from his analytical role as an NSA agent.

"That's alright Sam; I have had a big bunch of beliefs shaken lately. I understand how you feel and it's alright. I have several things to sort out myself."

JJ was thinking of the future, "If they have enough coordination to know that Suzanne was going to land at the Oakland airport, what are they going to do to us now when they know where we live, Owen?"

"I believe we can publicly spin you and your family out of this, to avoid the ire of the *Order* or any Aryan organizations. But we are also aware that the risk is around and in order to protect you, your family and Suzanne, I think we should temporarily relocate you in the meantime. Once our intel can assure that there isn't any buzz to retaliate against you personally, we can explore further options."

"So we can't go home?" Asked Karen.

"No, I'm afraid not. Look, this isn't an ideal situation from a family standpoint, but you do need to consider the risks. We will likely know if there are plans to retaliate against you, you have seen our resources. With the arrest of this new group, we can add to our knowledge the rest of their members to our data base. Not only can we listen, we can inject other reasons for them to have gotten caught. Believe me; the last thing I want is for you to be in any danger. Think of this time as a vacation, go to Europe. Suzanne would probably like to get to family; this will be a great opportunity for an extended stay."

Lester said, "If I may, I would like to go to my room. I have other work to do. I will see you tomorrow Karen, I always enjoy your presentations. This one should be a great one." He stood, looked at Owen and Sam, and said, "It's good to meet the faces of the people I never thought I'd meet. I look forward to not seeing you soon." He said in a humorous mood. In truth Lester was glad they turned out to be responsible agents. He imagined them twisting their guns on their index fingers, like Wild West cowboys. Knowing they seemed normal gave him another sense of relief.

Lester was already planning to come up with a method to use the power of deep meditation as a monitor for the potential use of this technology. He needed to get Hank and Karen together, once things settled. He needed some pregnant women, the net operating, and a control group of sterile women...

Lester was a thinker, designing experiments, and devising the controls and objectives. The lights, the visions of life Hank and other masters see; could it be the energy of the cosmic energy funneling into the brain cells. A sea of cosmic energy, a big pool that spills out radiation. They see it only sometimes... Lester saw the pool of energy and realized he tried to manage it all these years; and now was going to a role of really managing it, knowing what he was looking for. The pool manager; keeper of the détente.

92

At 11:45 p.m., five teams of FBI agents, six Siskiyou County Sheriffs, and three Dunsmuir policemen set up outside the west entrance of the compound described in the text from Johnstone to Deichmann. The communication between the FBI and the local police was productive in identifying the location and the best strategy for guarding the entrance and arresting Deichmann.

Deichmann and his female companion were followed from their airplane, photographed unloading several boxes into a rental car, and driving to the compound. The FAA's Flight Service was used to track Deichmann from Oakland North Field to Gillespie Field in El Cajon where he was met by the carpet van. After the hand-off, three agents followed Juergen and two other individuals who committed the break-in of the Jordan home.

The coordination of arrests went as planned and there was no gunfire in El Cajon, Deichmann's E 14th office, or in Dunsmuir where Roger and his companion were detained. After arresting him, the car was used by several agents to continue driving where intel had determined there were only four people in the compound; Johnstone among them. In a well coordinated assault, Johnstone and four other people were arrested and a cache of weapons, many computers, and loads of anti-Semitic materials were confiscated. If there were any communications to the outside world, it was not obvious; time would tell what the results of the bust would be.

07:54:22z—N.CA/FBI: Op Ordertake Dunsmuir Successful, Deichmann, Johnstone +6 in custody, no shots fired, equipment in custody.
07:55:00z—OAK PD: Op Ordertake Oakland Successful, no one on premises, evidence in custody.
07:59:23z—S.CA FBI: Op Ordertake El Cajon Successful, 3 in custody, no shots fired, evidence in custody. Further raids in progress no report yet.

Owen and Sam had left the hotel and were at the office. Owen was going to the NSA apartment, deep beneath the building designed for transient agents for a long sleep. Good. Nobody hurt. A home run.

The AD called a minute later and congratulated both agents for a job well done. He ordered them to sleep and take the day off. He suggested they go to the forum and watch the Jordan presentation, to make sure it could be kept mum, but go as citizens. That meant without the frame attached to their wrists. The technology was the greatest improvement in crime fighting since radios, but it was also draining. Keeping it on for too long could transfix you until the reality of it overtook the real reality; life, fishing, a movie...The AD knew these two agents needed to get away after the last week they had just lived.

"OK you're the boss, and thanks for giving me a couple of days. Maybe now you'll trust me all the time!" Laughed Owen. This was obviously a constant and running joke between the two men.

"I'll consider it next time, need one more point." Said the AD. That was apparently another running joke; it would always take one more right decision to allow Owen to have more latitude, a point that could not be given. Both agents knew the autonomous power with the technology available would be dangerous, the system required oversight.

"Night boss."

"Sam, make sure he gets some sleep and takes that damn wristband off."

"Yes Sir."

93

After the forum, Suzanne chose to return home to Croatia for an extended vacation. The Jordan's decided to spend the last non-school year with the boys in England. Lester studied Hinduism and practiced yoga, without the bending. Hank was helping Lester progress through the learning curve of meditation, Lester had no interest in it really, he was just curious if it could be taught.

Karen continued her research at a private home in the Cotswold's, a full house with plenty of equipment and security. Her healthy baby daughter was born on time with JJ present. They named her Stephanie. After time, she stopped looking over her shoulder for neo-Nazi's. JJ continued to travel, a new belief of what life is and a wary dose of the existence of evil that still exists.

Owen kept in touch with them all, reporting on anything of interest to the family and staying informed on the development of the modality. He said he had less belief in a God, but replaced it with a firm belief that if the organism is controlled by a set of rules, if they can be used to help cure disease, Dr. Karen Jordan could find the answer.

94

Epilogue

AP wire: Increasingly, in the former Yugoslavian country of Serbia, an abnormal number of babies are being born with what doctors are describing as Anencephaly. Anencephaly strikes an estimated one in 1,000 to one in 2,000 live births in the U.S., but this number has been decreasing due to better nutrition in developed countries. With early prenatal diagnosis, this disease typically leads to selective termination of the pregnancy. The increase of Anencephaly in the Serbian city of Apatin is alarming as the rate of babies with this rare disease has increased to one in 100, a factor increase above the average.

Apatin, Serbia is best known for its involvement in the guerrilla attacks into neighboring Croatia in the 1990's. Interviews with Croats in the city of Dalj, who were subjected to ethnic cleansing in the Serb-dominated areas, reported they had little or no sympathy for the affected Serbs. Government leaders and the National Health Coordinator in Apatin had few explanations, though a movement called the *Croats for Revenge* is taking credit for the increase, though this is widely reported to be a hoax.

World renown, Dr. Lester Warwick, a recognized specialist in brain studies, has been called to investigate this anomaly in association with the World Health Organization. So far there are no known environmental causes; officials are unaware of any reason for the increase in this rare affliction.